Fundamentals of Engineering Drawing

工程制图基础

主　编　张庆伟
副主编　夏　红　王建宏
参　编　易　于　薛寒松　朱小飞
主　审　宋主民　龚银玲

重庆大学出版社

内容提要

本教材是为配合大专院校工科基础课程——工程制图的双语教学而编写的。本书在不改变我国工程制图现有课程体系并遵照我国技术制图国家标准的前提下，参考国外同类教材，采用双语编写而成。其内容包括：制图的基础知识；点、直线和平面的投影；基本立体及表面交线；组合体；物体的表达；螺纹及螺纹连接件；零件图及装配图简介等，共 10 章。本书可供本科工科各专业教学用，也可供成人高校及网络远程教育教学使用。

图书在版编目(CIP)数据

工程制图基础/张庆伟主编.—重庆：重庆大学出版社，2007.9(2022.7重印)
ISBN 978-7-5624-3546-4

Ⅰ.工… Ⅱ.张… Ⅲ.工程制图 Ⅳ.TB23

中国版本图书馆 CIP 数据核字(2007)第 136779 号

FUNDAMENTALS OF ENGINEERING DRAWING
工程制图基础
Gongcheng Zhitu Jichu

主　编　张庆伟
副主编　夏　红　王建宏
参　编　易　于　薛寒松　朱小飞
主　审　宋主民　龚银铃

责任编辑：刘秀娟　贾兴文　刘丽萍　　版式设计：刘秀娟
责任校对：文　鹏　　　　　　　　　　　责任印制：赵　晟

*

重庆大学出版社出版发行
出版人：饶帮华
社址：重庆市沙坪坝区大学城西路 21 号
邮编：401331
电话：(023) 88617190　88617185(中小学)
传真：(023) 88617186　88617166
网址：http://www.cqup.com.cn
邮箱：fxk@cqup.com.cn (营销中心)
全国新华书店经销
POD：重庆新生代彩印技术有限公司

*

开本：890mm×1240mm　1/16　印张：17.25　字数：482 千
2007 年 9 月第 1 版　　2022 年 7 月第 7 次印刷
ISBN 978-7-5624-3546-4　定价：48.00 元

本书如有印刷、装订等质量问题，本社负责调换
版权所有，请勿擅自翻印和用本书
制作各类出版物及配套用书，违者必究

序

1. 课程的性质、任务和学习方法

图样和文字、数字一样,也是人类用以表达、构思、分析和交流思想的工具。在工程技术中,为了准确地表示出物体的形状、大小、材料等内容,通常将物体按一定的投影方法和技术规定表达在图纸上,称之为工程图样,简称工程图。工程图在工程技术上的应用非常广泛。无论是制造产品还是建造房屋,都必须先画出工程图,然后根据图纸加工或修建,才能得到预想的结果。因此,工程图被喻为工程语言。

本课程研究绘制和阅读工程图的原理和方法,培养学生的形象思维能力和空间思维能力,是一门既有系统理论又有较强实践性的技术基础课。近年来,计算机绘图技术突飞猛进,大大提高了绘图速度和质量,但是在应用计算机绘图技术之前,还必须掌握绘制工程图样的基本原理和方法。所以绘制和阅读工程图样的能力是工程技术人员必须具备的最基本的能力。

本课程的任务是:

1) 学习正投影法的基本原理和应用。
2) 培养对三维形状与相关位置的空间逻辑思维能力和形象思维能力。
3) 熟悉技术制图国家标准的有关规定,并学会查阅有关手册和相关的国家标准。
4) 培养绘制和阅读机械图样的能力。培养耐心细致的工作作风和严肃认真的工作态度。

在学习本课程时应注意以下几点:

1) 学习掌握正投影的基本原理及其应用时,应该坚持理论联系实际的学风,要认真学习投影原理。在理解基本概念、掌握正确作图方法的基础上,由浅入深地通过一系列的绘图和读图实践,不断地由物绘图,由图想物,分析和想象空间物体与图纸上图形之间的对应关系。以养成自觉地应用作图手段来构思、分析和表达工程问题的习惯。

2) 绘图时,必须遵守技术制图国家标准。在不断的绘图实践中养成自觉遵守国家标准的习惯。因为只有符合"国家标准"的图样才能相互交流。

3) 正确使用绘图工具和绘图仪器。所绘图样应该做到:投影正确、图线分明、尺寸齐全、字体工整、图面整洁。

要注意培养自学能力,要循序渐进地认真阅读课本,逐渐养成用英文思考的习惯。

2. 对教材的几点说明

本教材及配套习题集是为配合大专院校工科基础课程——工程制图课程的双语教学,在不改变我国工程制图现有课程体系并遵照我国技术制图国家标准的前提下,参考美国俄亥俄州立大学工程制图教授 THOMAS E. FRENCH 主编的教材《Engineering Drawing and Graphic Technology》和华盛顿大学机械工程教授 E. G. PARE 主编的教材《Descriptive Geometry》,用双语(主要用英文,个别不常见单词采用汉语注释)编写的工程制图教材。书中采用的几乎都是常见词汇,在写法上也尽量做到浅显易懂。本书内容包括:制图的基础知识;点、直线和平面的投影;基本立体及表面交线;组合体;物体的表达;螺纹及螺纹连接件、键、销、垫圈及齿轮;零件图及装配图简介等,共10章,可供本科工科各专业学生使用,也可供成人高校学生及网络远程教育学生使用。编写这套双语教材的目的是让学生在学习工程制图基础知识的同时,获得英语实践的机会,使学生把多年所学的英文作为工具来学习其他知识;同时,提高学习英文的兴趣,为今后进一步的学习和研究打下基础。

本书及习题集中的工程制图知识符合教育部2000年批准的《画法几何及工程制图课程教学基本要求》及《工程制图基础课程教学基本要求》。本教材和习题集所采用的技术制图标准符合国家质量技术监督局颁布的中华人民共和国国家标准。

本书第1章、4章、10章及附录由张庆伟编写;第2章由夏红编写;第3、5、7章由王建宏编写;第6章由朱小飞编写;第8章由薛寒松编写;第9章由易于编写;全书由张庆伟统稿和修改。重庆大学机械工程学院宋主民教授担任本教材的主审。重庆邮电大学外国语学院龚银玲老师担任英文指导并参与修改。

本书及配套习题集已入选了普通高等教育"十一五"国家级规划教材。在本书的编写过程中,我们得到重庆大学机械工程学院刘昌明教授、何玉林教授、丁一教授的指导和支持,同时还得到重庆大学国家工科机械基础课程教学基地杨学元、王喜庆老师的支持,在此一并表示感谢。

在本书的出版过程中,我们得到了重庆大学出版社编辑刘秀娟、刘丽萍的大力支持,在此深表感谢。

本书获重庆大学教材建设基金资助。

最后,竭诚欢迎广大读者对本书提出宝贵的意见和建议。

编 者

2007 年 8 月

Contents

Chapter 1　Basic Skills of Engineering Drawing ·· 1
　1.1　Drawing Instruments ·· 1
　1.2　Provisions of Chinese National Standard of Technical Drawing（机械制图国家标准的规定）·· 6
　1.3　Geometric Construction（几何构造）··· 14

Chapter 2　Points, Lines and Planes ·· 22
　2.1　Basic Theory of Projection ··· 22
　2.2　Principal Projection Planes（基本投影面）·· 25
　2.3　Projections of a Point ··· 26
　2.4　Views of a Line ··· 30
　2.5　Views of a Plane（平面的投影）··· 40

Chapter 3　Primary Objects ·· 51
　3.1　Polyhedra（多面体）·· 51
　3.2　Revolutions（回转体）··· 59

Chapter 4　Surface Intersections ·· 74
　4.1　Intersections of Planes and Polyhedra（平面与平面体的交线）····································· 74
　4.2　Intersections of Planes and Revolutions（平面与回转体的交线）·································· 80
　4.3　Intersections of Two Revolutionary Surfaces（两回转体的交线）································· 95

Chapter 5　Composite Objects ·· 105
　5.1　Projection Rules of an Object ··· 105
　5.2　Drawing Three Views ·· 106
　5.3　Dimensioning ·· 113
　5.4　Reading Views（读图）··· 119

Chapter 6　Views ··· 127
　6.1　Principal Views（基本视图）··· 127
　6.2　Removed Views（向视图）·· 130
　6.3　Partial Views（局部视图）··· 131

1

6.4　Auxiliary Views（辅助视图） ·· 132

Chapter 7　Sectional Views and Cross-Sectional Views ······················· 136
7.1　Sectional Views ·· 136
7.2　Types of Cutting-Planes ··· 146
7.3　Cross-Sections（断面图）·· 153
7.4　Other Representation Methods（其他表达方法）··· 157
7.5　Comprehensive Example ··· 164

Chapter 8　Threads, Fasteners and Gears ··· 166
8.1　Thread（螺纹）·· 166
8.2　Thread Fasteners and Stipulated Drawing（螺纹连接件及规定画法）·············· 181
8.3　Keys and Key Joining（键和键连接）··· 191
8.4　Pins（销）··· 194
8.5　Washers（垫片）·· 195
8.6　Gears（齿轮）··· 197

Chapter 9　Detail Drawings（零件图）·· 203
9.1　Contents ·· 203
9.2　Selecting Views ··· 204
9.3　Typical Parts ·· 205
9.4　Technical Requirements ·· 209
9.5　Reading Detail Drawings（阅读零件图）··· 221

Chapter 10　Introduction of Assembly Drawings ······································ 225
10.1　Contents ·· 225
10.2　Representations of Assembly Drawings ··· 225
10.3　Dimensions on An Assembly Drawing ··· 228
10.4　Part Numbered and Part List（零件序号和零件表）······································· 229

Appendix 1　Glossary（术语表）··· 230

Appendix 2 ··· 235

Chapter 1　Basic Skills of Engineering Drawing

A new machine, or product must exist in the mind of the engineer or designer before it becomes a reality. This original concept or idea is usually placed on paper called drawing paper to form engineering drawing. The engineers or designers can communicate each other with the engineering drawing. So, the engineering drawing is often compared to a working language.

Since drawing instruments are indispensable tools in making engineering drawings, this chapter will introduce some basic instruments used by engineers and drafters, and discuss the way to use them. Again, the Chinese National Standards of Technical Drawing are briefly introduced in the chapter. Lastly, some of geometric constructions are also brought in.

1.1　Drawing Instruments

1.1.1　Drawing pencils

A good drawing begins with a proper drawing pencil and its correct use. Pencil grade ranges from the hardest, 9H, to the softest, 7B. Grade H, HB and B are most frequently used because of their medium hardness. The choice of the proper drawing pencils depends on personal preference and the drawing paper used. Recently, mechanical pencils (自动铅笔) become popular. Leads in a mechanical pencil do not need sharpening but provide a uniform line.

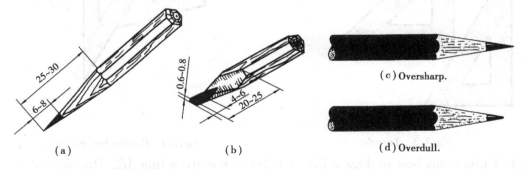

Fig. 1.1　Sharpen drawing pencils correctly.

If a wooden pencil is chosen, the procedure of sharpening it is as follows. First sharpen about 25 to 30 mm of the wood with a knife. Meanwhile, expose the lead about 6 to 8 mm. Note that the mark grade H or HB at the end of the pencil can not be sharpened. Second, a sandpaper is used to achieve the desired pencil point (笔尖) as shown in Fig. 1.1(a). You must exercise carefully in sharpening pencil point. Oversharp point may be broken or cut into the drawing paper easily (Fig. 1.1(c)). Conversely, overdull point, as shown in Fig. 1.1(d), will create fuzzy (模糊的) and inconsistent lines (不连续的线). The drawing pencil sharpened as shown in Figs.

1.1 (a) is used to draw thin line while the drawing pencil sharpened as shown in Fig. 1.1 (b) is used to brighten (加粗) thin line to form thick line.

1.1.2 Pencil eraser

A good eraser is necessary for removing the marks on paper quickly without smudging (污渍).

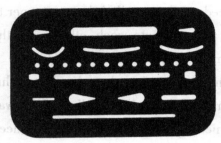

Fig. 1.2 Erasing shield.

1.1.3 Erasing shield (擦图片)

It is helpful to have an erasing shield similar to the type as shown in Fig. 1.2. The erasing shield will allow selective erasing without removing the drawings nearby. This can save time and enhance the quality of appearance of finished drawing.

1.1.4 Triangles (三角板)

Most inclined lines in engineering drawing are drawn with the 45° triangle and the 30°~60° triangle (Fig. 1.3). The triangles are made of transparent plastic (透明的塑料) so that lines of the drawing can be seen through them. With the help of a pair of triangles it is easy to generate lines parallel or perpendicular to a given line (Figs. 1.4 and 1.5).

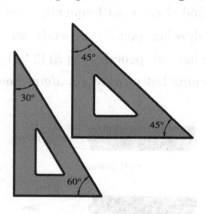

Fig. 1.3 Triangle.

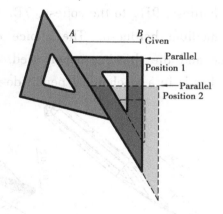

Fig. 1.4 Parallel line construction.

Fig. 1.4 illustrates how to draw a line parallel to the given line AB. One side of a triangle is placed along the given line AB first and then the supporting triangle is fixed against another side of the first triangle. Slide the first triangle along the supporting triangle to any position desired, and draw the parallel line.

Perpendicular lines may also be produced by the sliding triangle method. As shown in Fig. 1.5, one perpendicular side of a triangle is placed along the given line AB, the supporting triangle is then fixed against the hypotenuse (斜边) of the first triangle. Slide the first triangle across line AB along the supporting triangle to any perpendicular position desired, and draw the perpendicular line.

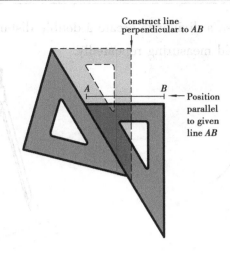

Fig. 1.5 Perpendicular line Construction.

1.1.5 Compass (圆规)

A compass is applied to draw circles and circular arcs (弧) (Fig. 1.6).

In order to obtain a high quality circle or circular arc, the lead of the compass must be properly sharpened and positioned correctly. A sharpening device such as a metal file (金属锉刀) or a sandpaper may be used to create the beveled side (斜面) of the lead as shown in Fig. 1.6 (a). The lead with beveled side is called compass lead (圆规芯).

To draw a circle, (1) set off (截取) the required radius on one of the center lines, (2) place the needle point at the exact intersection of the center lines, (3) adjust the compass to the required radius, and (4) lean (倾斜) the compass forward and draw the circle clockwise while rotating the handle between the thumb and forefinger (食指). To obtain a thick circle, it may be necessary to repeat the movement several times as shown in Fig. 1.6 (b).

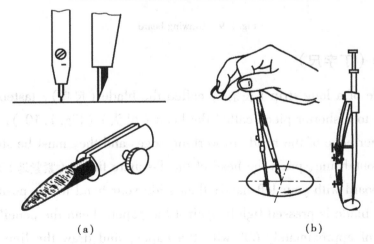

Fig. 1.6 Compass.

1.1.6 Divider (分规)

A divider is similar to a compass in construction. It is used for measuring distances from a ruler. It is also used for transferring distances or for setting off a series of equal distance (Fig. 1.7).

Fig. 1.8 illustrates how to use a divider to create a double distance simply by transferring the dimension measured, thus to avoid measuring repeatedly.

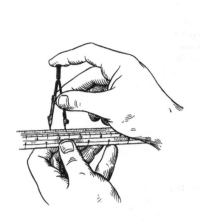

Fig. 1.7 Divider.

Fig. 1.8 Using divider.

1.1.7 Drawing board (图板)

A drawing board is made of wood on which the drawing paper is stuck (粘贴). The drawing board for student is shown in Fig. 1.9. The left side edge of the drawing board is working edge and it must be straight and smooth.

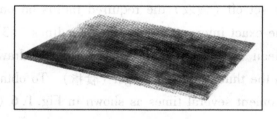

Fig. 1.9 Drawing board.

1.1.8 T-square (丁字尺)

T-square is made of a long strip (长条), called the blade (尺身), fastened rigidly (牢牢地固连) at right angles to a shorter piece called the head (尺头) (Fig. 1.10). The upper edge of the blade and the inner edge of the head are working edges and they must be straight and smooth.

To draw a horizontal line, press the head of the T-square firmly (紧紧地) against the working edge of the drawing board with your left hand; then slide your hand to the position shown in Fig. 1.11 (a) so that the blade is pressed tightly against the paper. Lean the pencil in the direction of the line at an angle of approximately 60° with the paper, and draw the line from left to right. Rotate the pencil about its axis while drawing so that its point will wear evenly (Fig. 1.11 (b)).

To draw a vertical line, press the triangle on the T-square with the vertical edge on the left, as shown in Fig. 1.12. With the left hand, press the head of the T-square against the drawing board; then slide the hand to the position shown in Fig. 1.12 where the hand must holds both the T-square and the triangle firmly in position. Draw the line upward, rotating the pencil slowly

Chapter 1 Basic Skills of Engineering Drawing

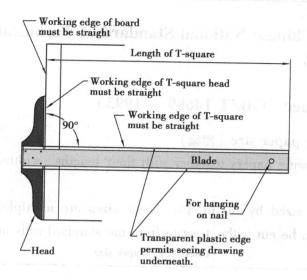

Fig. 1.10 T-square.

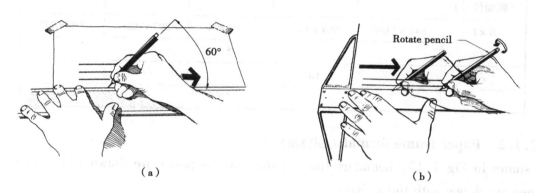

Fig. 1.11 Drawing a horizontal line.

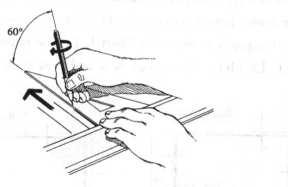

Fig. 1.12 Drawing a vertical line.

between the thumb and forefinger. Lean the pencil in the direction of the line at an angle of approximately 60° with the paper.

1.2 Provisions of Chinese National Standard of Technical Drawing（机械制图国家标准的规定）

1.2.1 Drawing paper（GB/T 14689—1993）

1.2.1.1 Drawing paper size（图幅）

The sizes of the drawing papers together with their lengths, widths and letter designations, are listed in Chart 1.1.

Drawing papers are sized by A0 to A4. These sizes are multiples of the standard size of 1188 × 841 and they can be cut without waste from the standard rolls of paper.

Chart 1.1　Paper size　　　　　　Unit（mm）

Format Code（幅面代号）	A0	A1	A2	A3	A4
$B \times L$	841 × 1189	594 × 841	420 × 594	297 × 420	210 × 297
a	25				
c	10			5	
e	20		10		

1.2.1.2 Paper frame format（图框格式）

As shown in Fig. 1.13, boundary lines of the drawing paper are drawn with thin line while frame lines are drawn with thick line.

Usually, there are two types of paper frames in use, i.e. without or with bookbinding（装订边）areas. The paper frame format without bookbinding area is shown in Fig. 1.13（a）. The value of e at each drawing paper is shown in Chart 1.1. The paper frame format with bookbinding area is shown in Fig. 1.13（b）. The values of a and c at each drawing paper are indicated in Chart 1.1.

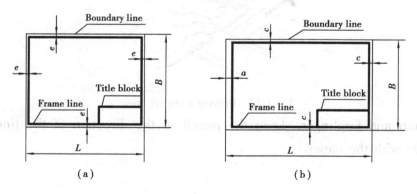

Fig. 1.13　Format of paper frame.

1.2.2 Title block（标题栏）

A title block is designed to show some information such as the name of the part, drawing scale, part material, part number, name of the drafter, name of the checker, name of the company, date of drawing and so on.

The title block is usually placed at the right bottom corner of the drawing paper (Fig. 1.13). However, sometimes drawing paper may be rotated 90° counterclockwise to make the title block be located at the right top corner of the drawing paper (Fig. 1.14).

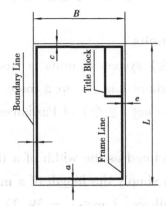

Fig. 1.14 Papers located vertically.

A typical title block is shown in Fig. 1.15 where all the words and digits should be written horizontally.

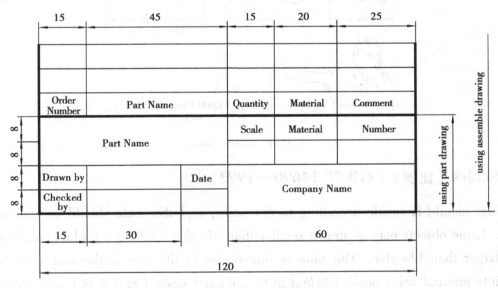

Fig. 1.15 Title block.

1.2.3 Measurement units for engineering(工程计量单位)

1.2.3.1 Metric system (SI) of units

The International Standard Organization (国际标准化组织, ISO) recommends to adopt the metric system for length in engineering, and the international system of units is abbreviated (缩写) to SI.

In the SI units, the meter is defined as a length equal to the distance traveled by light of a certain wavelength in a

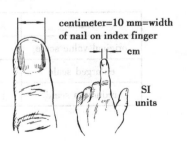

Fig. 1.16 The centimeter.

vacuum during a time interval of 1/299,792,458 second. In life, the nail width of your index finger (食指) is approximately equal to 1 centimeter, or 10 millimeters.

Besides, in engineering drawing, the units of all the dimensions are in millimeter (mm) which is one-thousandth of a meter, and its unit "mm" is generally omitted. For example, "25" means "25 mm".

1.2.3.2 English system of units

The English (Imperial 英帝国的) system of units is based on arbitrary units of inch, foot, cubit (腕尺), yard, and mile. England has set up a more accurate determination of the yard, which was legally defined in 1842 by act (法令) of Parliament (国会). A foot is 1/3 yard, and an inch is 1/36 yard.

In old England, an inch was defined as the width of a thumb or three barley corns (三棵麦粒), round and dry, and a foot was simply the length of a man's foot and so on (Fig. 1.17).

Conversion of two units is as follows: 1 meter = 39.37 inch, 1 inch = 25.4 mm

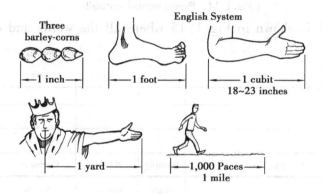

Fig. 1.17 English units.

1.2.4 Scales (比例) (GB/T 14690—1993)

Drawings should be made according to the scale, and the scale should be indicated in the title block. Large objects may be drawn smaller than life size (真实尺寸) while small objects may be drawn larger than life size. The ratio of drawn size to life size is the scale. Scales may be classified into original value scale (原值比例), enlarged scale (放大比例) and reduction scale (缩小比例) as shown in Chart 1.2. The scales in ( ) are second series and they are spare (备用) scales.

Chart 1.2 Scales

Kinds	Scale
original value scale	1:1
enlarged scale	2:1 (2.5:1) (4:1) 5:1 $1\times10^n:1$ $2\times10^n:1$ $5\times10^n:1$
reduction scale	(1:1.5) 1:2 (1:2.5) (1:3) (1:4) 1:5 $1:2\times10^n$ $1:5\times10^n$

Chapter 1 Basic Skills of Engineering Drawing

1.2.5 Characters（字体）（GB/T 14691—1993）

1.2.5.1 Height and number of a character（字高和字号）

Chinese National Standard of Technical Drawing stipulates that the height of a character may be 1.8, 2.5, 3.5, 5, 7, 10, 14 and 20 (unit: mm). Small letter（小写字母）h is used to assign the height of a character. The height of a character is defined as the number of the character. For example, the character of No. 3.5 means 3.5 mm in height.

1.2.5.2 Chinese characters（汉字）

In general, Chinese characters are square written symbols. In engineering drawing, Chinese characters are written in font of "长仿宋体"（Fig. 1.18）. If small letter h stands for the height of a Chinese character, its width is 0.7 h. As mentioned above, the Chinese character of No. 10 also means 10 mm in height.

No.10　字体工整笔画清楚间隔均匀排列整齐
No.7　横平竖直注意起落结构均匀填满方格
No.5　技术制图机械电子汽车航空船舶土木建筑矿山井坑满口纺织服装
No.3.5　螺纹齿轮端子接线飞行指导驾驶舱位挖填施工引水通风闸阀拥麻化纤

Fig. 1.18 Chinese characters in font of "长仿宋体".

Basic stroke in Chinese characters are point（点）, horizontal stroke（横）, vertical stroke（竖）, left-falling（撇）, right-falling（捺）, rising stroke（挑）, turning stroke（折）and hook stroke（钩）. The witten orders are shown in Chart 1.3.

Chart 1.3 Basic stroke in Chinese characters and the witten orders

名字	点	横	竖	撇	捺	挑	折	钩
基本笔划及运笔法	尖点 垂点 撇点 上挑点	平横 斜横	竖 直撇	平撇 斜撇	斜捺 平捺	平挑 斜挑	竖折 横折 撇点 横折折撇	竖钩 弯钩 平钩 横折钩 横折弯钩 右曲钩 竖弯钩 竖折折钩
举例	方光 心活	左七 下代	十千 上八	术分 建超	均公 技线	凹 安	周 及	牙子代买 孔力气码

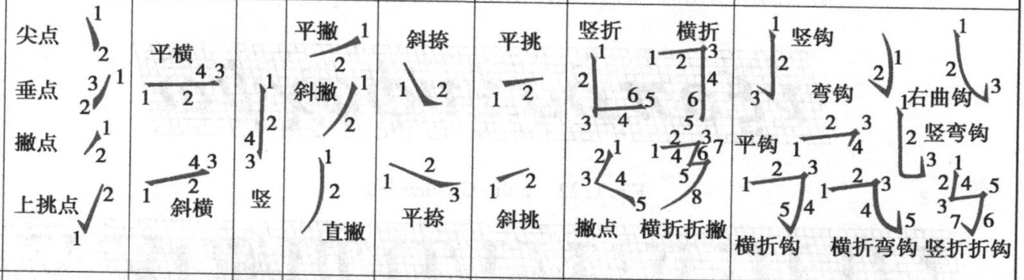

1.2.5.3 Letters and numbers（字母和数字）

Letters and numbers may be written in italic（斜体）or vertical（直体）. Italics are frequently applied in engineering drawing. Italic numbers are shown in Fig. 1.19; italic capitals in Fig. 1.20; italic small letters in Fig. 1.21; italic Grecian（希腊的）letters in Fig. 1.22 and Roman numerals in Fig. 1.23.

Fundamentals of Engineering Drawing

Fig. 1.19 Italic numbers.

Fig. 1.20 Italic capitals.

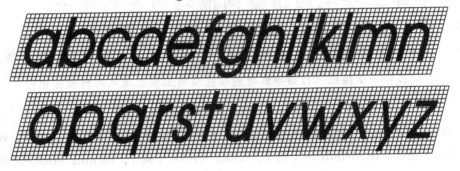

Fig. 1.21 Italic small letters.

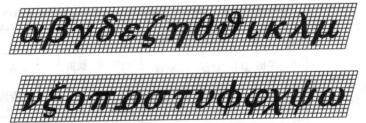

Fig. 1.22 Italic Grecian letters.

Fig. 1.23 Italic Roman numerals.

1.2.6 Drawing lines (图线) (GB/T1 7450—1998)

Each line on an engineering drawing has a definite meaning and is drawn in a certain way. Chart 1.4 shows various lines required in the engineering drawing.

Chart 1.4 Lines required in engineering drawing

Types	Weight	Example
Visible line（可见线）	b	
Hidden line（不可见线）	$b/3$	
Thin line（细线）	$b/3$	
Phantom line（假想线）	$b/3$	
Center line（中心线）	$b/3$	
Broken line (1) Wave line（波浪线）	$b/3$	
Broken line (2) long break line（长波浪线）	$b/3$	

Lines are classified into thick line and thin line. If letter b stands for the width of the thick line, then the width of thin line will be about $b/3$. In general, the b varies from 0.5 to 2 mm. For example, the width of thick line is 0.6 mm, thus the width of thin line is 0.2 mm.

The applications of all types of lines are shown in Fig. 1.24.

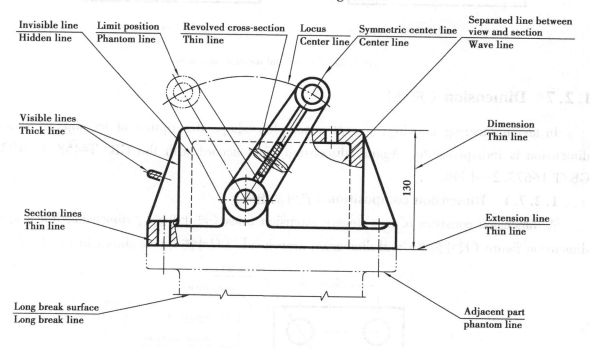

Fig. 1.24 Application of all types of lines.

Pay attention to several points:

When two hidden lines intersect, make sure the two short line segments intersect at crossing point, in order to avoid error 1 in Fig. 1.25 (b). It's the same when two center lines intersect, or a hidden line intersects a centerline. Avoid error 3 or error 5 in Fig. 1.25 (b).

When a centerline intersects an outline, make sure the two line segments intersect and the centerline segments should be 2 ~ 3 mm beyond the outline, neither too short nor too long. Avoid error 2, error 4 or error 6.

But when a hidden line is the extention of a thick line, the crossing point should be at interval. Avoid error 7 in Fig. 1.25 (b).

Besides, if a hidden line intersects a thick line, make sure the two line segments intersect to avoid error 8 in Fig. 1.25(b).

Fig. 1.25 (a) exhibits the correct way of drawing.

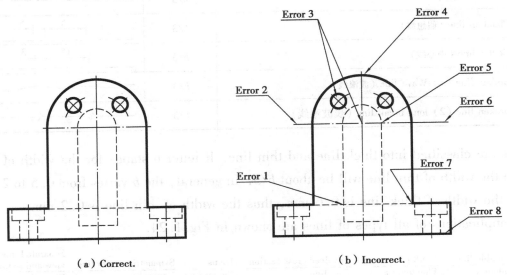

Fig. 1.25 Correct and incorrect drawings

1.2.7 Dimension (尺寸)

In an engineering drawing, except a complete shape description of an object, a complete dimension is indispensable. Again, dimension must accord with the GB/T4458.4—1984 and GB/T 16675.2—1996.

1.2.7.1 Dimension composition (尺寸的组成)

A dimension consistes of four items: extension lines (尺寸界线), dimension line (尺寸线), dimension figure (尺寸数字) and dimension arrowheads (尺寸箭头) as shown in Fig. 1.26.

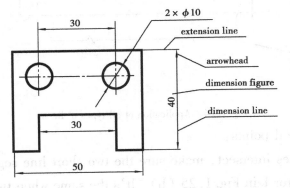

Fig. 1.26 Dimension terms.

A dimension line is a thin line terminated (终止) by arrowheads, which indicates the direction and extent (范围) of a dimension.

An extension line is a thin line that extends from a point on the drawing to which a dimension refers. The dimension line meets the extension lines at right angles. The extension line should extend about 3 mm beyond the outermost arrowhead. Besides, visible outlines or center line may be used for extension lines such as dimension 30 shown in Fig. 1.26.

Arrowheads indicate the extent of dimensions. They should be uniform in size and style throughout the drawing and not varied according to the size of the drawing or the length of dimensions. Arrowheads should be drawn freehand, and the length and width should be in a ratio of 4∶1. A larger sketch map (示意图) of arrowhead is shown in Fig. 1.27.

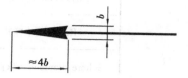

Fig. 1.27 Arrowheads.

A dimension figure is placed above a dimension line and it is written upward or to the left of the dimension line and it is written leftward.

In dimensioning, some errors are shown in Fig. 1.28 (b), while Fig. 1.28 (a) is correct.

Error 1. Arrowhead should contact with extension line.
Error 2. Dimension line can not be substituted by an outline.
Error 3. Dimension line can not be substituted by a center line.
Error 4. Dimension "ϕ" can not be substituted by dimension "R".
Error 5. Dimension line should be perpendicular to extension lines.
Error 6. Dimension figure's direction and position are incorrect.
Error 7. Arrowhead is incorrect.
Error 8. Dimension figure can not be traversed by any line.
Error 9. The shorter dimension should be the nearest to the object outline.

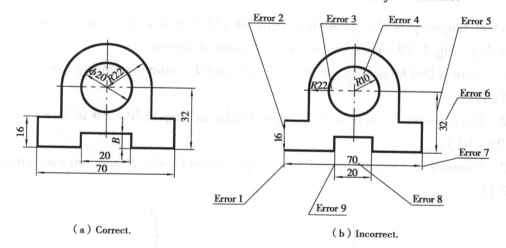

Fig. 1.28 Correct and incorrect in dimensioning.

1.2.7.2 Common dimension symbols (Chart 1.5)

Common dimension symbols used in engineering drawing are listed in Chart 1.5.

Chart 1.5 Common dimension symbols

Name	Symbol	Name	Symbol
Diameter（直径）	φ	Square（正方形）	□
Radius（半径）	R	45° Chamfer（倒角）	c
Sphere diameter（球直径）	Sφ	Depth（深度）	↧
Sphere radius（球半径）	SR	Socket（沉孔）	⊔
Arc（弧）	⌒	Countersink（埋头孔）	∨
Thickness（厚度）	T	Uniformity（均布）	EQS

1.3 Geometric Construction（几何构造）

Simple geometrical structures should be familiar to engineers for they occur frequently in engineering drawing. For example, how to draw a tangent arc? or a hexagon? Geometrical graphics are two-dimensional figures which are composed of straight lines, curves, or the combination of them. Geometric constructions in this chapter include tangent arcs, regular polygons, ellipses and so on.

1.3.1 How to draw tangent points

A point of tangency is the theoretical point at which a line joins an arc or two arcs join without crossing. Fig. 1.29 shows how to find the point of tangency.

Step 1 Draw a line tangent to the circle with center E, which is tangent line AB as shown in Fig. 1.29 (a).

Step 2 From center E, draw a line perpendicular to tangent line AB to locate tangent point T (Fig. 1.29 (b)).

Step 3 Connect two centers E and D to locate tangent point P between two tangent circles (Fig. 1.29 (b)).

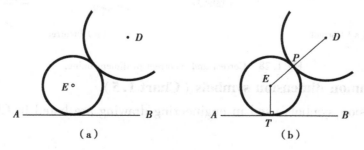

Fig. 1.29 Tangent points.

1.3.2 How to draw a line tangent to a circle through a point on the given circle

In Fig. 1.30 (a), given the arc *ABC*, draw a line tangent to the given circle at point *C*.

Step 1 Arrange a triangle in combination with the T-square (or another triangle) so that its hypotenuse passes through center *O* and point *C*.

Step 2 Holding the T-square firmly in place, turn the triangle about its square corner and move it until the hypotenuse passes through *C*.

Step 3 Draw the tangent line required along the hypotenuse (Fig. 1.30 (b)).

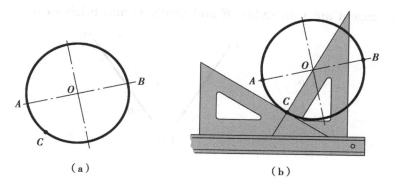

Fig. 1.30 A line tangent to a circle through a point on a circle.

1.3.3 How to draw a line tangent to a circle through a point outside the given circle

As shown in Fig. 1.31 (a), the arc *ABC* and point *P* outside the arc *ABC* are given, and then draw a tangent line at the point *P*.

Step 1 Connect points *O* and *P*, and then locate point *D* by bisecting *OP* (Fig. 1.31 (b)).

Step 2 With point *D* as center and *OD* as the radius, draw a semicircle to intersect the given arc tangent point at *T*, (Fig. 1.31 (c)).

Step 3 Draw the line from *P* to *T*, which is the tangent line required.

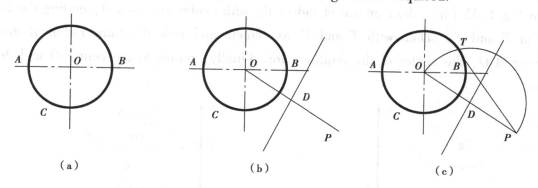

Fig. 1.31 A line tangent to a circle through a point outside the circle.

1.3.4 How to draw an arc tangent to two lines

In Fig. 1.32 (a), if lines *AB*, *CD* and radius *R* are given, draw tangent arc between the given lines. Drawing procedure is shown in Fig. 1.32 (b).

Step 1　Set compass to the given radius R, and at any convenient point on the given lines draw many arcs.

Step 2　Through the limits of those arcs, draw parallels to the given lines, which are the loci (轨迹, locus 的复数) of centers of all circles tangent to lines AB and CD, and the intersecting point O is the center of the tangent arc.

Step 3　Find two tangent points T by erecting perpendiculars to the given lines through the center O.

Step 4　Draw tangent arc with radius R and center O and brighten it.

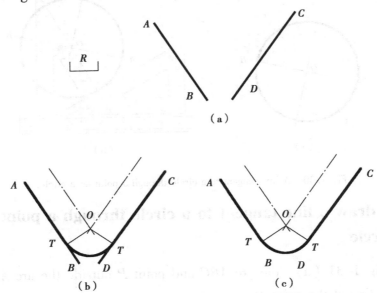

Fig. 1.32　An arc tangent to two straight lines.

1.3.5　How do draw an arc tangent to two perpendicular lines

Fig. 1.33 (a) is similar to Fig. 1.32 (a), but line AB is perpendicular to line AC. As shown in Fig. 1.33 (b), draw an arc of radius R, with center at corner A, cutting the lines AB and AC at T and T_1. Then with T and T_1 as centers and with the same radius R draw arcs intersecting at O, the center of the required arc. Finally, return to the center O and draw the same arc.

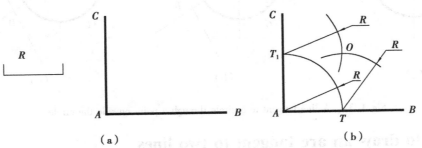

Fig. 1.33　An arc tangent at right-angle corner.

1.3.6 How to draw an arc tangent to a circle and a line

In Fig. 1.34 (a), line AB is given, and arc with radius R_1 and center O is given. Draw a tangent arc with radius R between the given line and arc.

Step 1 Draw a line CD parallel to AB at a distance R from it.

Step 2 With O as center and $R+R_1$ as radius, swing an arc intersecting CD at X, which is the desired center of the tangent arc.

Step 3 Draw lines XT and OX to locate the tangent points (Fig. 1.34 (b)).

Step 4 Draw arc with center X and radius R and brighten it (Fig. 1.34 (c)).

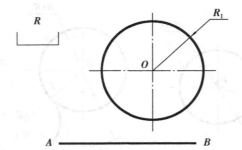

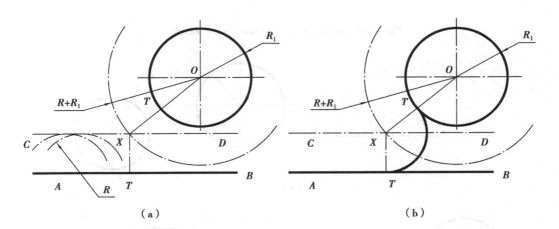

(a)　　　　　　　　　　　　　　(b)

Fig. 1.34 An arc tangent to a straight line and a circle.

1.3.7 How to draw an arc tangent to two circles

An arc tangent to the two given arcs has two cases. The first case is that the centers of the given circles are outside the tangent arc. In Fig. 1.35 (a), two arcs are given: one is with radius R_1 and center O; the other is with radius R_2 and center P. Draw a tangent arc with radius R between the given arcs.

Step 1 With O as center and $R + R_1$ as radius, draw an arc (Fig. 1.35(b)).

Step 2 With P as center and $R + R_2$ as radius, swing another arc intersecting the first arc at Q, which is the center sought.

Step 3 Find out the tangent points T on lines OQ and QP.

Step 4　Draw the tangent arc with radius R and center O and brighten it (Fig. 1.35 (c)).

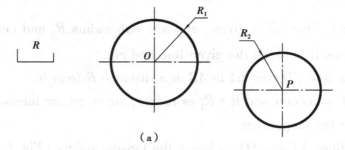

(a)

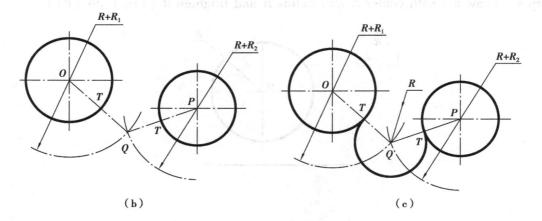

(b)　　　　　　　　　　　(c)

Fig. 1.35　An arc tangent to two circles outside.

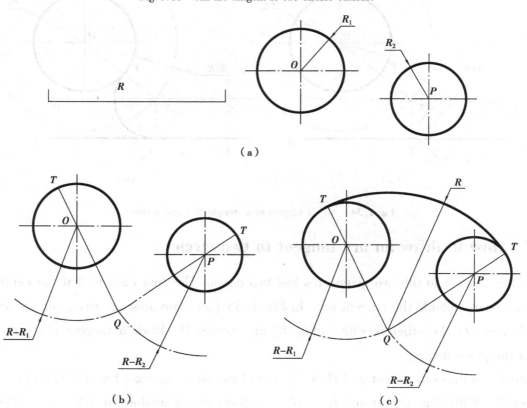

(a)

(b)　　　　　　　　　　　(c)

Fig. 1.36　An arc tangent to two circles inside.

The second case is that centers of the given circles are inside the tangent arc. In Fig. 1.36 (a), the given circles are the same with Fig. 1.36(a) while the radius of tangent arc becomes larger. Draw tangent arc between the given circles.

Step 1 With O as center and $R - R_1$ as radius, draw an arc (Fig. 1.36 (b)).

Step 2 With P as center and $R - R_2$ as radius, swing another arc intersecting the first arc at Q, which is the center sought.

Step 3 Find out the tangent points T in the extended lines of OQ and QP.

Step 4 Draw the tangent arc with radius R and center O and brighten it (Fig. 1.36 (c)).

1.3.8 Hexagons (六边形)

The hexagon, a six-side regular polygon, can be inscribed in (圆内接) or circumscribed (圆外切) about a circle.

Use a 30°~60° triangle to draw the inscribed hexagon and circumscribed hexagon (Figs. 1.37(a) and (b)). The circle represents the distance from corner to corner for an inscribed hexagon and from flat to flat (边到边) for a circumscribed hexagon.

Fig. 1.37 (c) shows another method to construct the inscribed hexagon with a compass and straightedge.

Draw a circle with AB as a diameter. With the same radius and corners A and B as center respectively, draw two arcs to intersect the circle at points 3, 4, 5, 6 and connect the adjacent points with straightedge to form the regular polygon.

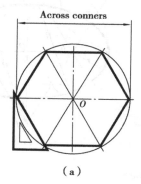
(a)

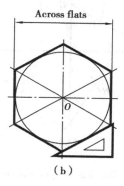
(b)

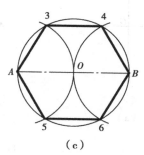
(c)

Fig. 1.37 A hexagon.

1.3.9 Pentagons (五边形)

The pentagon is a five-side regular polygon. Fig. 1.39 (a) shows how to draw an inscribed pentagon.

Step 1 Bisect radius OB to locate point D.

Step 2 With D as center and DC as radius, draw arc CE to locate E.

Step 3 With C as center and CE as radius draw arc EF to locate F.

Step 4 Use line CF as the chord (弦) to locate the other corners of the pentagon. If five diagonals (对角线) are connected respectively, a pentagram (五角星) can be formed as shown in

Fig. 1.38 (b).

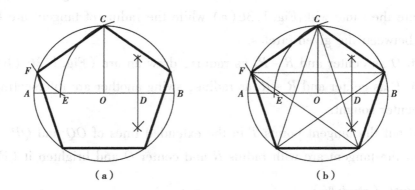

Fig. 1.38 A regular pentagon in a circle.

1.3.10 How to draw a ellipse

Ellipse is one type of conic curves (圆锥曲线). The largest diameter of an ellipse is called major axis (长轴) while the shortest diameter is called minor axis (短轴). Both axes intersect perpendicularly and bisect each other. There are two methods to construct ellipse.

1.3.10.1 Approximate circular-arc method (近似圆弧法)

The approximate circular-arc method is shown in Fig. 1.39. Axes AB and CD are given in Fig. 1.39 (a).

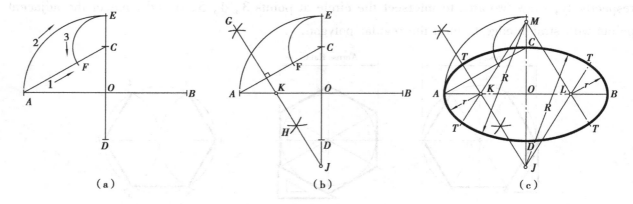

Fig. 1.39 Drawing an approximate ellipse.

Step 1 Draw line AC. With O as center and OA as radius, draw arc AE. With C as center and CE as radius, draw arc EF (Fig. 1.39 (a)).

Step 2 Bisect line segment AF to obtain bisector GH which intersects both axes at K and J, the centers of the required arcs (Fig. 1.39 (b)).

Step 3 Find centers M and L by setting off (截取) $OL = OK$ and $OM = OJ$.

Step 4 Tangent points are located on the lines joining the centers, which are line JK, JL and CK, CL. As shown in Fig. 1.39 (c), R is the radius of large arc and r is the radius of the small arc.

Step 5 Draw two large arcs with radius R and center C, J, respectively.

Step 6 Draw two small arcs with radius r and center K, L, respectively.

The four arcs constitute the required ellipse.

1.3.10.2 Concentric-circle method (同心圆法)

The major axis AB and minor axis CD of an ellipse are given. Construct the ellipse, using the concentric-circle method (Fig. 1.40).

Step 1   Construct two concentric circles with the center O and radii OA and OD, respectively.

Step 2   From a number of points on the outer circle, as P and Q, draw radii, OP, OQ, etc. intersecting the inner circle at P', Q', etc.

Step 3   From P and Q draw lines parallel to OD, and from P' and Q' draw lines parallel to OB. The intersection of the lines through P and P' gives one point on the ellipse, the intersection of the lines through Q and Q' another point, and so on.

Step 4   The process above may be repeated in each of the four quadrants and then connect all the points to form the ellipse.

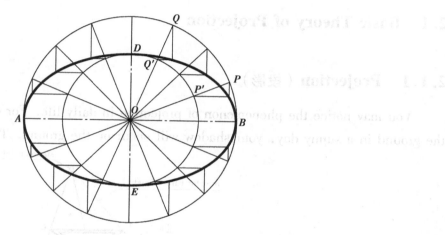

Fig. 1.40   An ellipse by concentric-circle method.

Chapter 2　Points, Lines and Planes

The knowledge and skills we will study in this chapter are fundamentals of making engineering drawings. It involves the projection（投影）of three-dimensional objects（三维物体）on the two-dimensional plane（二维平面）. Through proper geometric maniputions, projections may reflect the object's lengths, angles, shapes and other geometric information. The projection of the object may be considered as the view（视图）. The main tasks of this chapter are to study three views of points, lines and planes, respectively.

2.1　Basic Theory of Projection

2.1.1　Projection（投影）

You may notice the phenomenon of projection in daily life. For example, when you stand on the ground in a sunny day, your shadow will occur on the ground. The shadow is a projection.

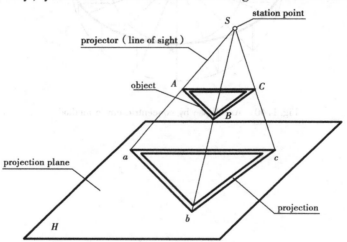

Fig. 2.1　Perspective projection.

A projection of an object is actully an image（图像）produced by projecting（投射）the object onto a plane known as the projection plane（投影面）. As shown is Fig. 2.1, suppose that point S is a light source and from it, many light lines called projectors（投射线）are projected out. Suppose that again if an object, triangle ABC, is projected by these projectors, an image—triangle abc, would occur on the projection plane H. The triangle abc is the projection of the object, triangl ABC. How to make the projection abc? The projector through corner point A intersects（交）projection plane H at a, Which is the projection of point A. If the same procedure is applied to corner points B and C, projections b and c can be obtained. When projections a, b, c are connected by straight lines, projection abc is obtained.

Besides, the light source may be called station point (投射中心) or observer, and projectors may be known as lines of sight. In general, capital letters are assigned to spatial points and small letters stand for projection points.

2.1.2 Classification of projection

According to the relative position between projectors in space, two types of projections are defined, which are perspective projection (透视投影) and parallel projection (平行投影), respectively.

2.1.2.1 Perspective projection

A perspective projection is a projection produced by projecting an object onto the projection plane under those projectors which are converged to the station point. Perspective projection is not suitable for working drawings because a perspective projection view does not reveal the exact size and shape. It is used to some extent by engineers in preparing preliminary (初步的) sketches.

2.1.2.2 Parallel projection

If the distance from the observer to an object is infinite far away, all projectors are parallel to each other. So the projection produced by projectors parallel to each other is called parallel projection (Fig. 2.2). According to the relative position between projectors and projection plane, the parallel projection can be classified further into two types: oblique projection (斜投影) and orthographic projection (正投影).

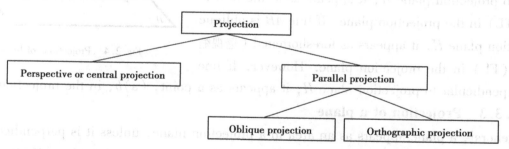

In oblique projection, projectors are oblique to the projection plane while in orthographic projection, projectors are perpendicular (垂直) to the projection plane. Orthographic projection can represent the true size and shape of an object, again the drawing procedure is simple, thus it is widely used in engineering drawings. In this book, projection refers to the orthographic projection, unless otherwise specified.

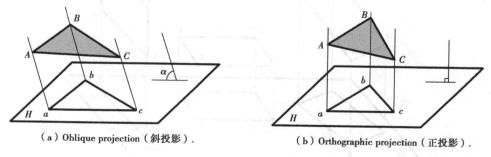

(a) Oblique projection (斜投影). (b) Orthographic projection (正投影).

Fig. 2.2 Parallel projections.

2.1.3 Characteristics (特性) of orthographic projection

2.1.3.1 Projection of a point

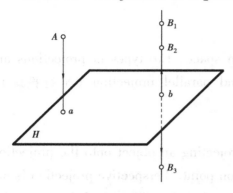

Fig. 2.3 Projections of points.

For a spatial point, its projection is unique in a projection plane. On the contrary, the spatial position of a point can not be determined solely by its one projection.

In Fig. 2.3, the projector from point A is perpendicular to plane H and the foot of the perpendicular (垂足) is unique. So, the projection of point A in the plane H is unique and it is labeled by small letter a. However, the projection of any point on the projector Bb is always projection b, so the spatial position of point B can not be identified by projection b only.

2.1.3.2 Projection of a line

A line appears as a line in a projection plane generally, unless it is perpendicular to the projection plane. In orthographic projection, a projection of a line can not be longer than itself. In Fig. 2.4, if line AB is parallel to projection plane H, it appears as a line in true length (TL) in the projection plane. If line AB is oblique to projection plane H, it appears as foreshortened (透视缩短) line (FL) in the projection plane. However, If line AB is perpendicular to projection plane H, it appears as a point, $(a)b$, in the projection plane.

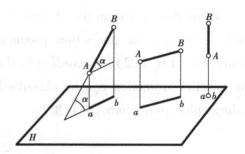

Fig. 2.4 Projections of lines.

2.1.3.3 Projection of a plane

In general, a plane appears as an area in a projection plane, unless it is perpendicular to the projection plane. In Fig. 2.5, if plane $ABCD$ is parallel to the projection plane H, it appears as an area in true size (TS) in the projection plane. If plane $ABCD$ is oblique to projection plane H, it appears as an area in foreshortened surface (FS) in the projection plane. If plane $ABCD$ is

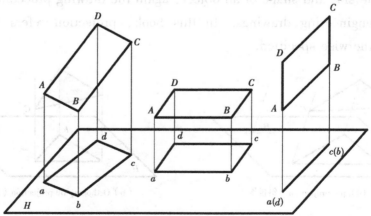

Fig. 2.5 Projections of planes.

perpendicular to the projection plane H, it appears as an edge in the projection plane.

2.2 Principal Projection Planes (基本投影面)

Since it is impossible to determine the position of a point with its only one projection, more projection planes are added.

Usually, three projection planes perpendicular to each other are used in orthographic projection (Fig. 2.6(a)). They are horizontal projection plane (水平投影面), frontal projection plane (正立投影面) and profile projection plane (侧立投影面), denoted by H, V and W, respectively.

Intersection lines of any two adjacent (邻近的) projection planes are defined as projection axes. The intersection line of plane V and plane H is defined as axis X; the intersection line of plane H and plane W as axis Y; the intersection line of plane V and plane W as axis Z. Axes X, Y, and Z are perpendicular to each other in space and converge at point O known as projection origin (投影原点).

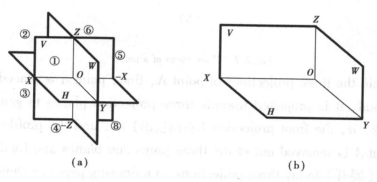

Fig. 2.6 Principal projection planes.

In Fig. 2.6(a), three projection planes divide space into eight parts or quadrants (象限) numbered from 1 to 8. According to the Chinese National Standard (国家标准) of Technical Drawings, the first-angle projection (第一角投影) is used to make engineering drawings while in some other countries, such as in the U.S and Canada, the third-angle projection (第三角投影) is used in technical drawings. Hence, this textbook is focused on the first-angle projection. Note that in first-angle projection, an object is placed in quadrant ①, and observer always looks through the object towards the projection plane. But in third-angle projection, the object is placed in quadrant ③, and observer always looks through the projection plane towards the object. In third-angle projection, projection plane is assumed transparent (透明的). In summary, three projection planes in the first-angle make up of a projection system in our country, which is called three projection planes system (三面投影体系).

2.3 Projections of a Point

2.3.1 Rules of projections of a point(点的投影规律)

Theoretically, a point has location but no dimensions(尺寸). For the sake of accuracy, a point is indicated by a tiny circle, as shown in Fig. 2.7, other than a dot.

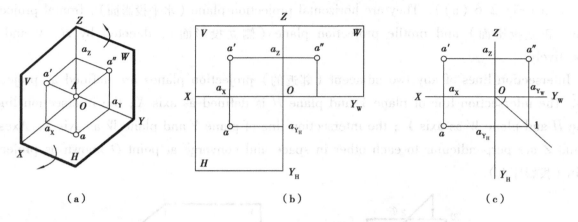

Fig. 2.7 Three views of a point.

In order to obtain the three projections of point A, first, point A is placed in three projection planes system. Second, it is projected towards three projection planes to generate the horizontal projection(水平投影) a, the front projection(正面投影) a', and the profile projection(侧面投影) a''. Third, point A is removed out of the three projection planes and finally, three projection planes are unfolded(展开) to lay three projections on a drawing paper as shown in Fig. 2.7 (b).

In unfolding, assume that plane H and plane W are hinged(铰接) to plane V and keep plane V stationary, each projection plane revolves outward from original position until it lies in plane V, that is, drawing paper. Note that in revolving, axis Y is divided into two parts, Y_H and Y_W as shown in Fig. 2.7 (b).

From plane geometry, it is known that a plane has no bounds, thus borderlines of projection plane are omitted as shown in Fig. 2.7 (c).

A projection of an object is known technically as a view. In other words, frontal projection, horizontal projection and profile projection are front view(主视图), top view(俯视图) and left view(左视图), respectively.

Evidently, the position of the point A can be determined by its three views a, a', a'' or at least by any two views.

Geometry knowledge tells us that two intersecting lines make a plane. So, projectors Aa, Aa' construct a plane which intersects axis OX at a_X. In the same way, a_Y and a_Z are obtained. When the plane H is folded up, line aa_X will become vertical and line up with a_Xa'. Thus, $a'a_X$ and a_Xa form a single straight line $a'a$ which is perpendicular to projection axis X. Similarly, when the plane W is folded up, line $a''a_Z$ will become horizontal and line up with a_Za'. Thus, $a''a_Z$ and

$a_Z a'$ form a single straight line $a''a'$ which is perpendicular to projection axis Z. Based on above analysis, projection rules of a point can be summarized as follows:

1) Projection line, aa', is perpendicular to the axis OX ($aa' \perp OX$).

2) Projection line, $a'a''$, is perpendicular to the axis OZ ($a'a'' \perp OZ$).

3) The distance aa_X is equal to the distance $a''a_Z$ because both of them represent the distance from the spatial point A to plane V ($aa_X = a''a_Z$).

The projection rules of a point tell us that through one projection point, we can and can only draw two projection lines which are perpendicular to projection axis, respectively. Through a', two projection lines can be drawn: one is perpendicular to axis OX—$a'a$ and the other one is perpendicular to axis OZ — $a'a''$. Through a'', two projection lines can be drawn: one is perpendicular to axis OZ — $a''a'$ and the other one is perpendicular to axis OY_W, i. e. $a''a_{Y_W}$. Through a, two projection lines also can be drawn: one is perpendicular to axis OX, i. e. aa' and the other one is perpendicular to axis OY_H, i. e. aa_{Y_H}.

As shown in Fig. 2.7(c), when projection lines aa_{Y_H} and $a''a_{Y_W}$ are extended, they intersect at point 1. If point O and point 1 are connected to form one line which is the diagonal of the square $oa_{Y_H} 1 a_{Y_W} O$ and it is called the 45° miter-line.

It is the 45° miter-line that relates the top projection point a and the profile projection point a''. In other words, the distance aa_X and $a_Z a''$ can be transferred to each other by the 45° miter-line.

According to the projection rules, the third projection of a point can be worked out by any two projections.

Example 2.1 Given a' and a as the front and top views of point A; b and b'' as the top and left views of point B, work out the third views of points A and B (Fig. 2.8(a)).

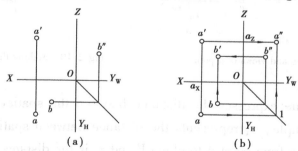

Fig. 2.8 The third views of points A and B.

Analysis is as follows.

The given view shows that the third view of point A is the left view a'' while the third view of point B is the front view b.

Graphic solution is as follows.

First, the left view a'' is obtained by a' and a.

Step 1 Through a, draw its projection line to intersect the 45° miter-line at point 1.

Step 2 Through point 1, draw the perpendicular line of axis OY_W to meet the other projection line drawn from a' at a''.

In the drawing procedure above, the distance aa_X is transferred to the profile view to form the distance $a''a_Z$ by means of the 45° miter-line.

Second, based on b and b'', b can be obtained.

Through b, draw its projection line to the front view to intersect the other projection line drawn from b'' at b'.

2.3.2 Relationship between views of a point and its coordinates (点的投影与其坐标间的关系)

The three projection axes can be regarded as coordinates axes to form a coordinates system. The projection origin O is the origin of the coordinates system. Any point can be determined in space by its coordinates x, y and z.

In Fig. 2.9, spatial point A is determined either by its three coordinates x_A, y_A, z_A or by its three views, a', a, a''. That is, $A(x_A, y_A, z_A)$ or $A(a', a, a'')$ is a determined point in space. The relation between coordinates and the views are shown in Fig. 2.9. That is,

$$x_A = Oa_X = a_Y a = a_Z a' = a''A$$
$$y_A = Oa_Y = a_Z a'' = a_X a = a'A$$
$$z_A = Oa_Z = a_X a' = a_Y a'' = aA$$

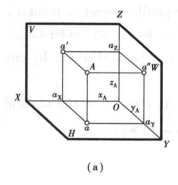
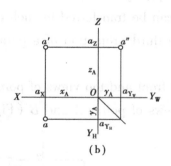
(a)　　　　　　　　(b)

Fig. 2.9 Coordinates and views of a point.

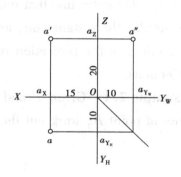

Fig. 2.10 Making the three views of the given point A.

Each coordinates value expresses the distance between the spatial point and corresponding projection plane. For example, x_A represents the distance between spatial point A and plane W. Similarly, y_A is the distance from point A to plane V and z_A is the distance from point A to plane H.

In addition, any view of a point is determined by the two coordinates of the point. For example, the top view a is determined by x_A and y_A, while the front view a' by x_A and z_A and the left view a'' by y_A and z_A.

Example 2.2 Make the three views of the given point $A(15, 10, 20)$. See Fig. 2.10.

Analysis is as follows.

Measure coordinates along axes to obtain a_x, a_Y, a_Z first and then through them draw projection lines. The intersection point of per two projection lines is a view of the point. Note that measuring reference point in three directions, X, Y and Z, is the origin O.

Graphic solution is as follows.

Step 1　From origin O, measure $Oa_x = x_A = 15$ on axis X; $Oa_{Y_H} = y_A = Oa_{Y_W} = 10$ on axis

Y, and $Oa_Z = z_A = 20$ on axis Z.

Step 2　Through a_X, a_{Y_H}, a_{Y_W} and a_Z draw projection lines, respectively.

Step 3　The intersecting point of per two projection lines is the views a', a, and a'', respectively.

2.3.3　Relative position (相对位置) of two points

A point may be said to be above or below, in front of or behind, or to the left or to the right of another point. These spatial relative positions are described in the views of the two points on the same projection plane.

In Fig. 2.11, if given $\triangle x = 20$, $\triangle y = 10$ and $\triangle z = 15$, this means that point B is 20 mm to the right of point A, 15 mm above point A, and 10 mm behind point A.

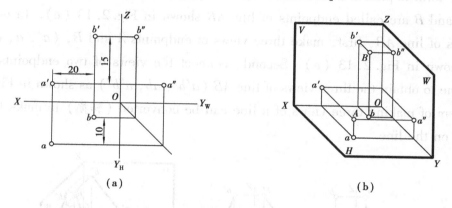

Fig. 2.11　Relative position of two points.

2.3.4　coincident points (重影点) and their visibility

Those points located on one projector are called coincident points. When projecting along projector, they appear as one projection point. In other words, those points coincide in a certain projection plane.

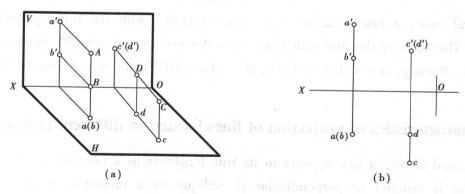

Fig. 2.12　Coincident points and visibility.

The visibility of projection of coincident points can be judged by comparing their coordinates' values. Point with larger value of coordinates is always visible.

In Fig. 2.12 (a), because point A and point B are located on one projector of plane H, they are a pair of coincident points on plane H. When point A and point B are projected to plane H, their top views coincide as one projection point. In addition, because point A is above point B, in top view, point A is visible and point B is invisible. Invisible projection point is labeled by ( ).

Similarly, points C and D are coincident points on plane V.

2.4 Views of a Line

2.4.1 Views of a line

The term line is generally used to designate a straight line unless otherwise specified. Theoretically, a line is of indefinite length but frequently when the conditions of the problem make it obvious, the term implies a definite line segment (线段).

Points A and B are called endpoints of line AB shown in Fig. 2.13 (c). In order to obtain the three views of line AB, first, make three views of endpoints A and B, (a', a, a'') and (b', b, b''), as shown in Fig. 2.13 (a). Second, connect the views of two endpoints in the same projection plane to obtain the three views of line AB ($a'b'$, ab, $a''b''$) as shown in Fig. 2.13 (b). So, the problem of making three views of a line can be converted (转换) to make three views of two endpoints on the line.

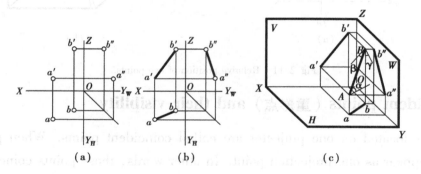

Fig. 2.13 Views of a line.

In general case, a line is at an acute angle (锐角) with the three projection planes, respectively. The angle of the line with plane H is denoted by α, and with plane V by β, with plane W by γ. Namely, α = (Line $AB\char`\^H$), $\beta$ = (Line $AB\char`\^V$) and γ = (Line $AB\char`\^W$). See Fig. 2.13 (c).

2.4.2 Characteristics of projection of lines located in different positions in space

As discussed above, a line appears in its true length or as a point or a foreshortened line when the line is parallel or perpendicular or oblique to a projection plane, respectively. Generally, a view of a line cannot be longer than the line itself in orthographic projection. Lines located in different positions will be introduced as follows.

2.4.2.1 Principal lines(投影面的平行线)

A principal line is a line that is parallel to one of the three projection planes but oblique to the other two. Such line may be parallel to plane V (frontal line (正平线)), to plane H (horizontal line (水平线)) or to plane W (profile line (侧平线)). This line appears in its true length on the projection plane to which it is parallel and foreshortened length on adjacent projection planes. The true-length view of a princcipal line is always oblique to projection axes, while the foreshortened-length views are always parallel to projection axes. True angles, α, β, γ, appear on the true-length view.

A horizontal line lies in or parallel to plane H as shown in Chart 2.1 Row 1.

A frontal line lies in or parallel to plane V as shown in Chart 2.1 Row 2.

A profile line lies in or parallel to plane W as shown in Chart 2.1 Row 3.

Chart 2.1 Principal lines

Horizontal line (水平线) ∵ Line AB // plane H ∴ $Z_A = Z_B$			①Line AB appears in its true length in the top view, that is $ab = AB$. ②$\beta = (ab \wedge OX) = (AB \wedge V)$, $\gamma = (ab \wedge OY_H) = (AB \wedge W)$. ③Line AB appears as foreshortened length in the front and left views respectively and $a'b'$ // OX, $a''b''$ // OY_W.
Frontal Line (正平线) ∵ Line AB // plane V ∴ $Y_A = Y_B$			①Line AB appears in its true length in the front view, that is, $a'b' = AB$. ②$\alpha = (a'b' \wedge OX) = (AB \wedge H)$, $\gamma = (a'b' \wedge OZ) = (AB \wedge W)$. ③Line AB appears in foreshortened length in the top and left view respectively and ab // OX, $a''b''$ // OZ.
Profile Line (侧平线) ∵ Line AB // plane W ∴ $X_A = X_B$			①Line AB appears as true length in the left view that is, $a''b'' = AB$. ②$\alpha = (a''b'' \wedge OY_W) = (AB, H)$, $\beta = (a''b'' \wedge OZ) = (AB \wedge V)$. ③Line AB appears in foreshortened length in the front and top views respectively and $a'b'$ // OZ, ab // OY_W.

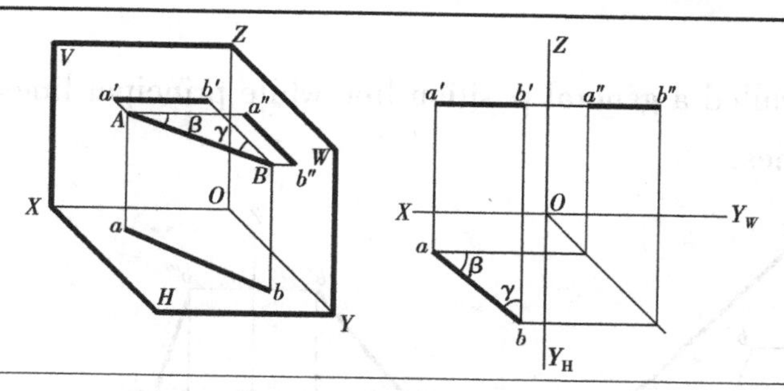

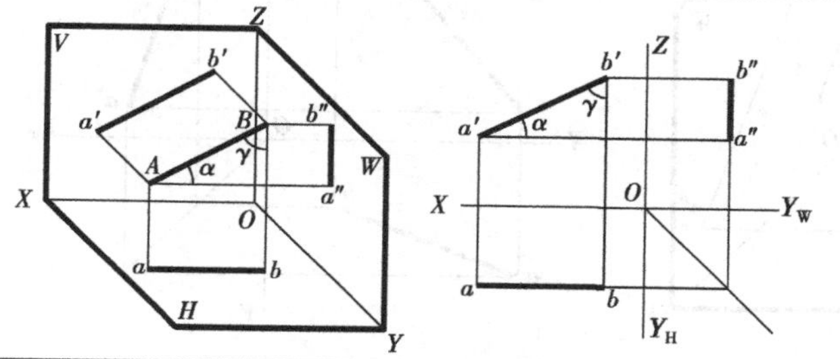

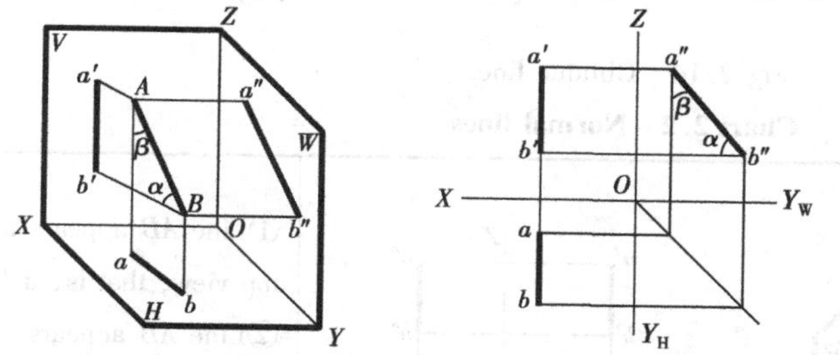

2.4.2.2 Normal lines(投影面的垂直线)

A normal line is a line that is perpendicular to one of the three projection planes but parallel to the other two. Such line may be perpendicular to plane V(正垂线), to plane H(铅垂线) or to plane W (侧垂线). This line appears as a point on the projection plane to which it is perpendicular and as a line in its true length on adjacent projection planes.

A normal line of plane H is perpendicular to plane H and parallel to plane V and plane W as shown in Chart 2.2 Row 1.

A normal line of plane V is perpendicular to plane V and parallel to plane H and plane W as shown in Chart 2.2 Row 2.

A normal line of plane W is perpendicular to plane W and parallel to plane V and plane H as shown in Chart 2.2 Row 3.

2.4.2.3 Oblique lines (倾斜线)

An oblique line is a line that is oblique to all the projection planes. Since it is not perpendicular to any projection plane, it cannot converge into a point in any view. Since it is not parallel to any projection plane, it cannot appear in its true length in any view. Thus an oblique line always appears in foreshortened length and in an inclined position in all three views (Fig.2.14 (b)).

Angle between a spatial line and three projection planes are denoted by α, β and γ, respectively (Fig.2.14 (a)). Since there is no true length in any view, true angle cannot appear in any view.

In addition, an oblique line is called a general position line while principal lines and normal lines are called particular position lines.

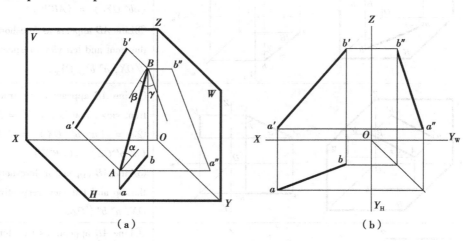

Fig. 2.14 Oblique line.

Chart 2.2 Normal lines

A normal line of plane H (铅垂线) ∵ Line $AB \perp$ plane H ∴ $X_A = X_B$ $Y_A = Y_B$			①Line AB appears as a point in the top view, that is, a (b). ②Line AB appears in its true length in the front and left views, that is, $a'b' = AB$, $a''b'' = AB$. ③$a'b' // OZ$, $a''b'' // OZ$.
A normal line of plane V (正垂线) ∵ Line $AB \perp$ plane V ∴ $X_A = X_B$ $Z_A = Z_B$			①Line AB appears as a point in the front view, that is, a' (b'). ②Line AB appears in its true length in the top and left views, that is, $ab = AB$, $a''b'' = AB$. ③$ab // OY_H$, $a''b'' // OY_W$.

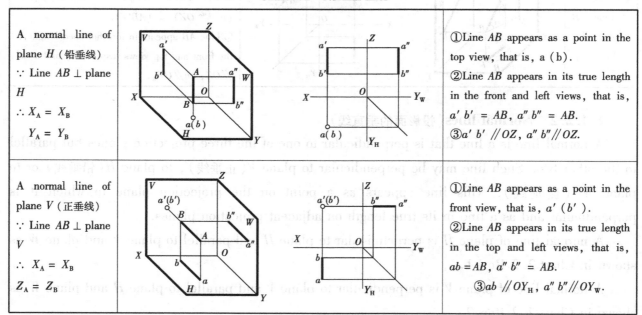

A normal line of plane W (侧垂线) $\because$ Line $AB \perp$ plane W $\therefore Y_A = Y_B$ $\quad Z_A = Z_B$	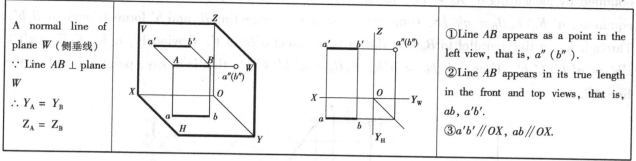		①Line AB appears as a point in the left view, that is, $a''(b'')$. ②Line AB appears in its true length in the front and top views, that is, ab, $a'b'$. ③$a'b' /\!/ OX$, $ab /\!/ OX$.

2.4.3 Points on lines

With respect to points on lines, two projection characteristics are as follows.

1) If a point is on a line in space, the views of the point appear on the corresponding views of the line. Any two adjacent views of the point must lie on a projection line.

2) Points dividing a line segment in a given ratio will divide any view of the line in the same ratio.

In Fig. 2.15 (a), the three views of line AB are ab, $a'b'$ and $a''b''$. If point K is on line AB, its top view k must be on ab; its front view k' must be on $a'b'$; its left view k'' must be on $a''b''$. Again, any two adjacent views of point K have relations of $k'k \perp OX$, $k'k'' \perp OZ$. Evidently, equation $AK:KB = ak:kb = a'k':k'b' = a''k'':k''b''$ comes into existence (成立). Fig. 2.15 (b) shows the unfolded three views of a point on a line.

According to the given views, how to judge whether a point is on a line or not? In general case, if a line is an oblique line and any two views of a point appear on the corresponding views of the oblique line, the point must be on the line in space.

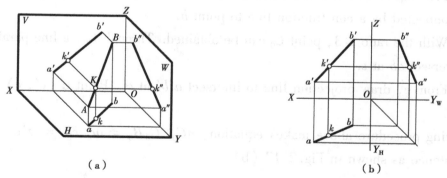

Fig. 2.15 A point on a line.

However, the exception occurs when a line is a principal line. In Fig. 2.16 (a), the front and top views show that line AB ($a'b'$, ab) is a profile line. Although k' belongs to $a'b'$ and k belongs to ab, it can not be determined whether point K is on line AB or not. An additional view is drawn, in which k'' does not belong to $a''b''$. So, point K is not on the line AB (Fig. 2.16 (b)). Again, the same conclusion can also be obtained by constructing two auxiliary similar triangles. In Fig. 2.16 (c), with a' as reference point, draw any auxiliary line $a'T_0$. Points K_0 and B_0 can be obtained by measuring $a'K_0 = ak$, $K_0B_0 = kb$ on the auxiliary line $a'T_0$. In other words, the equation, $a'K_0 : K_0B_0 = ak : kb$, come into existence. Connecting B_0 and b' forms triangle $a'B_0b'$. Through K_0, a line parallel to B_0b' is drawn to intersect $a'b'$ at t'. Obviously, $a'K_0 : K_0B_0 = a't' : t'b' \neq a'k' : k'b'$. Thus, $ak : kb = a'K_0 : K_0B_0 \neq a'k' : k'b'$, that is to say, point K is not on the line AB.

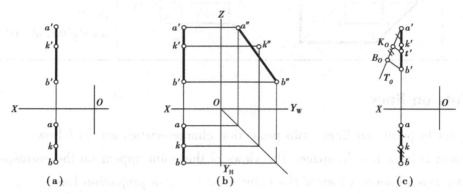

Fig. 2.16 Determining whether a point is on a line.

Example 2.3 Given $a'b'$ and ab as two views of line AB as shown in Fig. 2.17 (a), locate point C on the line AB by means of $AC : CB = 2 : 3$.

Analysis is as follows.

To solve the problem is actually to find the two views of the division point C on the line AB. Using the given ratio, constructing two similar triangles obtains one view of the division point C. The other view of point C can be obtained by means of the property of a point on the line.

Graphic solution is as follows.

Step 1 In the top view, with a as reference point, any auxiliary line is drawn.

Step 2 Five equal divisions are marked on the auxiliary line and the fifth division point, point B_0, is connected by a construction line to point b.

Step 3 With the ratio $2:3$, point C_0 can be obtained. Through C_0, a line parallel to line B_0b is drawn to intersect ab at c.

Step 4 From c, draw projection line to intersect $a'b'$ at c'. Point C (c', c) is the division point required.

The drawing procedure above makes equation, $aC_0 : C_0B_0 = ac : cb = a'c' : c'b' = 2 : 3$, come into existence as shown in Fig. 2.17 (b).

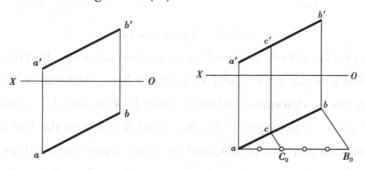

Fig. 2.17 Dividing a line segment in a given ratio.

2.4.4 Relative positions of two lines

There are three kinds of relative positions of two lines including intersecting, parallel, and skew.

2.4.4.1 Intersecting lines

Intersecting lines are lines that contain a common point (共有点). Any two adjacent views of the common point must lie on a single projection line perpendicular to projection axis.

In Fig. 2.18 (a), lines AB and CD intersect at point K which is a common point of the two lines. This fact is reflected in the three views as shown in Fig. 2.18 (b). From the projection characteristics of points on the line, it is known that the three views of point K appear on the corresponding views of lines AB and CD, respectively. Namely, k' belongs to both $a'b'$ and $c'd'$. In other words, $a'b'$ intersects $c'd'$ at k'. Similarly, ab intersects cd at k, and $a''b''$ intersects $c''d''$ at k''. Again, the front view k' and top view k must lie on a projection line. Similarly, the front view k' and left view k'' must lie on a projection line ($k'k \perp OX$, $k'k'' \perp OZ$).

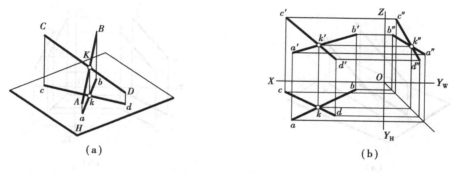

Fig. 2.18 Intersecting lines.

On the contrary, if all views of two lines intersect respectively, and any adjacent views of intersecting point lie on a projection line, the two lines shown by the given views intersect in space.

For example, in Fig. 2.19, two lines appear as intersecting lines in their three views,

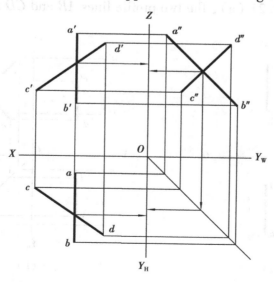

Fig. 2.19 Determining two skew lines.

respectively. But, it is evident that apparent-crossing-points (the term is explained in 2.4.4.3) of any two adjacent views do not lie on a projection line. So, the two lines are not intersecting lines.

2.4.4.2 Parallel lines

Parallelism of lines is a property that is preserved in orthographic projections. Thus lines parallel in space project as parallel lines in any view except in those views in which they coincide or appear as points—situations that do not alter the fact that the lines are parallel.

In Fig. 2.20 (a), lines $AB /\!/ CD$, projectors through all the points on lines AB and CD establish two parallel planes which intersect projection plane H at ab and cd. From plane geometry, it is known that ab and cd must be parallel (Two parallel planes intersect another plane to generate intersecting lines which are parallel.). In the same reason, $a'b' /\!/ c'd'$ and $a''b'' /\!/ c''d''$. The three views of two parallel lines are shown in Fig. 2.20 (b).

Conversely, if the lines in the views are respectively parallel, the lines themselves are parallel in space.

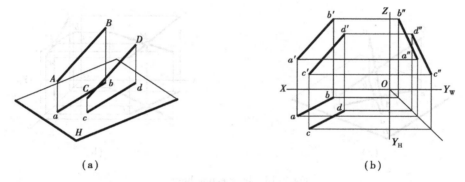

Fig. 2.20 Parallel lines.

However, two horizontal, two frontal or two profile lines that appear to be parallel in two views may or may not be actually parallel in space.

For example, in Fig. 2.21 (a), the two profile lines AB and CD appear parallel in their front

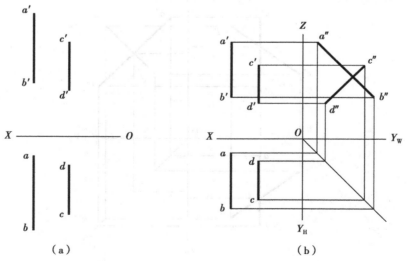

Fig. 2.21 Check of parallelism of principal lines.

and top views. Without further study, it might be concluded that the lines are parallel in space; but when the left view is added (Fig. 2.20 (b)), it is apparent that the two lines are not parallel.

2.4.4.3 Skew lines(交叉直线)

Skew lines are lines that are neither intersecting nor parallel. Skew lines are always in different planes, respectively as shown in Fig. 2.22 (a), but any two lines in a plane must either intersect or be parallel.

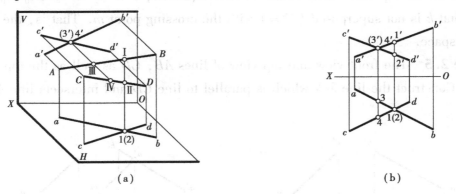

Fig. 2.22 Skew lines.

Since the lines do not intersect, there is no common point on the lines. The crossing point in the view is the projection of a pair of coincided points and it is called apparent-crossing-points (表观交点). In Fig. 2.22 (b), the apparent-crossing-points in the top view are projected to the front view, resulting in two separate points, $1'$ and $2'$. It is observed that $1'$ on $a'b'$ is higher than $2'$ on $c'd'$, so the apparent-crossing-points appear as 1 (2) in the top view.

Similarly, the apparent-crossing-points in the front view are projected to the top view, resulting in two separate points, 4 and 3. It is observed that 4 on cd is in front of 3 on ab, so, the apparent-crossing-points appear as $4'$ $(3')$ in the front view.

Example 2.4 As shown in Fig. 2.23, judge the relative position of two lines.

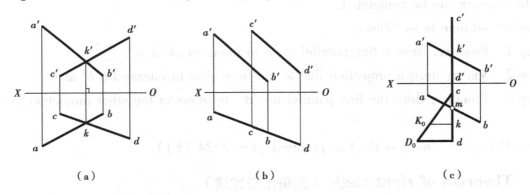

Fig. 2.23 Judge the relative position of two lines.

In figure (a), lines AB and CD appear as intersecting lines in their front and top views, and again the front and top views of the crossing point lie on a single projection line perpendicular to projection axis. So, the two lines intersect in space.

In figure (b), lines AB and CD appear as parallel in the front view while as co-line (共线)

in the top view, which implies the plane represented by two parallel lines is perpendicular to plane H. So, the two lines are parallel to each other in space.

In figure (c), although lines AB and CD appear as intersecting lines in their front and top views, it can not be determined whether the two lines AB and CD intersect actually in space. Right conclusion can be made by adding the third view or drawing two similar triangles. In figure (c), k' is assigned the crossing point in the front view and m is the crossing point in the top view. Locate k on the top view cd by means of ratio equation, $c'k' : k'd' = cK_0 : K_0D_0 = ck : kd$. It is evident that k is not superposed (重合) with the crossing point m. That is, the two lines are skew lines in space.

Example 2.5 The front view and top view of lines AB, CD as well as the top view of point M are given. Construct the line MN which is parallel to line CD and intersects line AB at point N (Fig. 2.24).

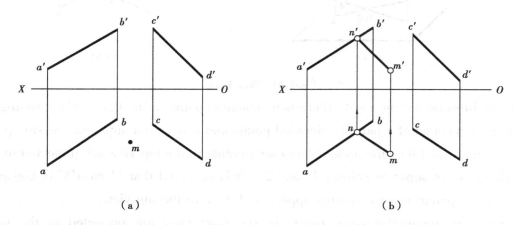

Fig. 2.24 Example 2.5.

Analysis is as follows.

By means of the projection principle of points on lines and parallelism property of two parallel lines, the drawing can be completed.

Graphic solution is as follows.

Step 1　From m, draw a line parallel to cd to intersect ab at n.

Step 2　Via n, draw a projection line to the front view to intersect $a'b'$ at n'.

Step 3　From n', draw the line parallel to $c'd'$ to intersect the other projection line from m at m'.

Line MN ($n'm'$, nm) is the line required (Fig. 2.24 (b)).

2.4.5　Theorem of right angle (直角投影定理)

In descriptive geometry, the theorem of projection of right angles is stated thus: If two lines are perpendicular, they appear perpendicular in any view showing at least one of the lines in true length. Conversely, if two lines appear perpendicular in a view, they are actually perpendicular in space only if at least one of the lines is true length in that same view.

For example, pictorial drawing in Fig. 2.25 (a) shows that lines AB and BC are

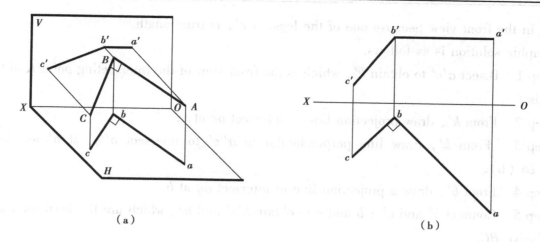

Fig. 2.25 Theorem of projection of right angle.

perpendicular and one of lines, line AB, is parallel to plane H. The true 90° angle still appears in the top view because one of the legs, ab, is true length.

The theorem is proved as follows:

Given $AB \perp BC$ and $AB /\!/ H$, prove $ab \perp bc$.

$\because AB /\!/ H$, $Bb \perp H$ (Bb is a projector of plane H.), $\therefore AB \perp Bb$.

$\because AB \perp BC$, $\therefore AB \perp$ plane $BbcC$ (If a line is perpendicular to two intersecting lines, the line is perpendicular to the plane represented by the two intersecting lines). $\therefore AB \perp bc$ (If a line is perpendicular to a plane, it is perpendicular to every line in the plane).

Again, $\because AB /\!/ H$, $\therefore AB /\!/ ab$, that is, $ab \perp bc$.

Fig. 2.25 (b) shows unfolding theorem of projection of right angles in the front and top views.

Example 2.6 In Fig. 2.26 (a), line AC is a diagonal (对角线) of a rhombus (菱形). Edge AB is one of four edges and it is located on line AQ. Complete views of the rhombus.

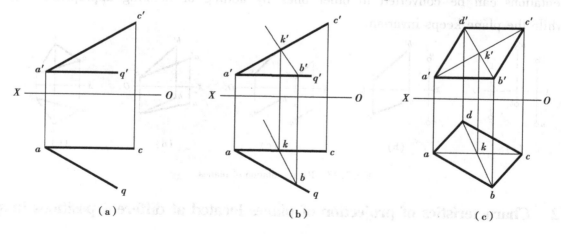

Fig. 2.26 Complete views of the rhombus.

Analysis is as follows.

Plane geometry tells us that the two diagonal of a rhombus intersect perpendicularly and the intersecting point bisects two diagonals. The two views in Fig. 2.26 (a) show that line AC is parallel to plane V. By means of theorem of projection of right angles, the true 90° angle still appears in the front view because one of the legs, $a'c'$, is true length.

Graphic solution is as follows.

Step 1 Bisect $a'c'$ to obtain k', which is the front view of the intersecting point K of the two diagonals.

Step 2 From k', draw projection line to intersect ac at k.

Step 3 From k', draw line perpendicular to $a'c'$ to intersect $a'q'$ at b' as shown in Fig. 2.26 (b).

Step 4 From b', draw a projection line to intersect aq at b.

Step 5 Connect b' and c'; b and c to obtain $b'c'$ and bc, which are the front view and top view of edge BC.

Step 6 By means of parallelism property, complete the two views of the rhombus as shown in Fig. 2.26 (c).

2.5 Views of a Plane (平面的投影)

A plane is a surface such that a straight line connecing any two points in that surface lies wholly within the surface. The general term "plane" implies a plane indefinite in extent unless other specified, such as a rectangle. Any two lines in a plane must either intersect or be parallel.

2.5.1 Representation of planes

A plane may be uniquely represented by three points not in a straight line (Fig. 2.27(a)). A plane may also be represented by a line and a point out of the line, or by two parallel lines, or by two intersecting lines or by a plane figure as shown in Fig. 2.27 (b) to (e). Each of these representations can be converted to other ones by adding or deleting appropriate lines and meanwhile the plane keeps invariant.

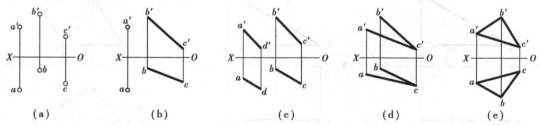

Fig. 2.27 Representation of planes.

2.5.2 Characteristics of projection of planes located at different positions in space

As discussed before, a plane may appear in its true size surface (TS) or as foreshortened surfaces (FS) or an edge when the plane is parallel or perpendicular or oblique to a projection plane, respectively. According to the relative positions of a plane to projection planes, the plane may be oblique plane, or principal planes (投影面的平行面), or inclined planes (投影面的垂直面). Three letters α, β and γ are used to indicate the angles between the spatial plane and three projection planes. The angle of the plane to plane H is α, to plane V is β and to plane W is γ.

2.5.2.1 Inclined planes(投影面的垂直面)

An inclined plane is a plane that is perpendicular to one of the three projection planes but inclined to the other two. Such plane may be perpendicular to plane V (正垂面) or to plane H (铅垂面) or to plane W (侧垂面). This plane appears as an edge inclined to projection axes on the projection plane to which it is perpendicular and as foreshortened surface on adjacent projection planes.

Chart 2.3 shows the projection characteristics of these planes.

Chart 2.3 Characteristics of projection of inclined planes

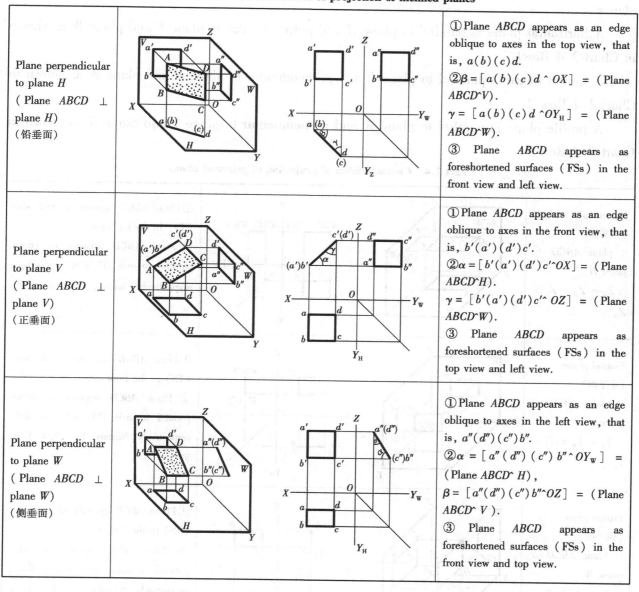

It should be pointed out that a spatial plane and its FS or two FSs must accord with the geometric relationship as follows:

①They have the same number of edges. For example, FS of a triangle must be a triangle. It is impossible that the FS of the triangle becomes a rectangle.

②The parallelism between edges does not change. For example, subtenses (对边) of a rectangle is parallel to each other. So, the subtenses in FS of the rectangle should be parallel, but the angle between adjacent sides and the length of sides are changed in the FS.

2.5.2.2 Principal planes(投影面的平行面)

A principal plane is a plane that is parallel to one of the three projection planes and perpendicular to the other two. Such plane may be parallel to plane V, (frontal plane 正平面), or to plane H, (horizontal plane 水平面), or to plane W, (profile plane 侧平面). This plane appears in its true size on the plane to which it is parallel, and as edges on adjacent projection planes.

A horizontal plane is parallel to plane H and perpendicular to plane V and plane W as shown in Chart 2.4 Row1.

A frontal plane is parallel to plane V and perpendicular to plane H and plane W as shown in Chart 2.4 Row 2.

A profile plane is parallel to plane W and perpendicular to plane V and plane H as shown in Chart 2.4 Row 3.

Chart 2.4 Characteristics of projection of principal planes

Horizontal plane (水平面) ∵ plane $ABCD$ // plane H ∴ $Z_A = Z_B = Z_C = Z_D$			①Plane $ABCD$ appears as true size (TS) in the top view. ② Plane $ABCD$ appears as edges parallel to axis OX and axis OY_W, respectively. Namely, $Z_A = Z_B = Z_C = Z_D$.
Frontal plane (正平面) ∵ plane $ABCD$ // plane V ∴ $Y_A = Y_B = Y_C = Y_D$			①Plane $ABCD$ appears as true size (TS) in the front view. ② Plane $ABCD$ appears as edges parallel to axis OX and axis OZ, respectively. Namely, $Y_A = Y_B = Y_C = Y_D$.
Profile plane (侧平面) ∵ Plane $ABCD$ // plane W ∴ $X_A = X_B = X_C = X_D$			①Plane $ABCD$ appears as true size (TS) in the left view. ② Plane $ABCD$ appears as edges parallel to axis OZ and axis OY_H, respectively. Namely, $X_A = X_B = X_C = X_D$.

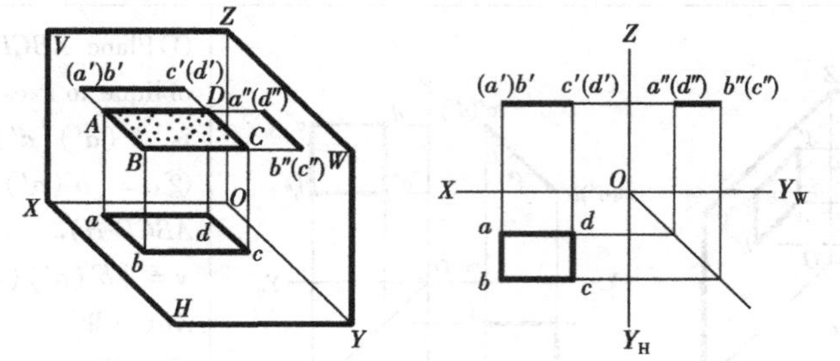

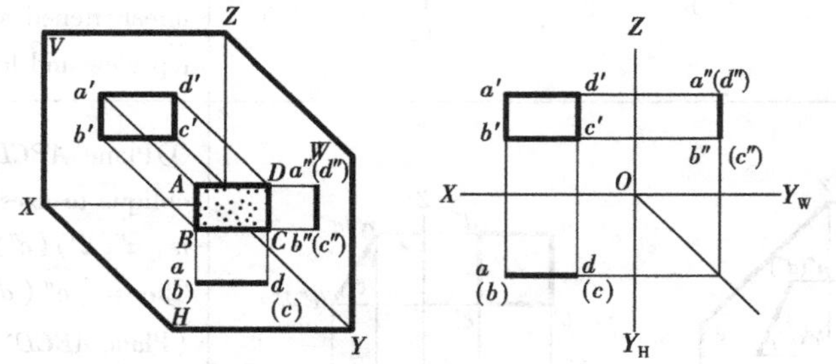

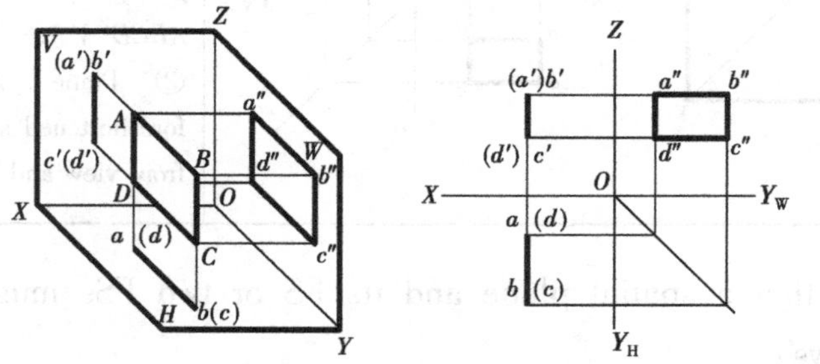

2.5.2.3 Oblique planes (倾斜面)

An oblique plane is a plane that is oblique to all the projection planes. Since it is not perpendicular to any projection plane, it cannot appear as an edge in any view. Since it is not parallel to any projection plane, it cannot appear in its true size surface in any view. Thus an oblique plane always appears as a foreshortened surface in all three views (Fig. 2.28).

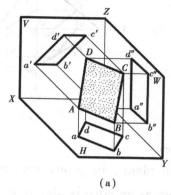 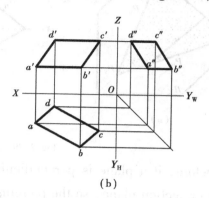

(a)　　　　　　　　　　　　　　(b)

Fig. 2.28　Oblique planes.

2.5.3　Representation of planes—piercing line method (迹线表示法)

Suppose that a plane is extended to pierce (穿过) the three projection planes to generate intersections which are called piercing lines (迹线). In Fig. 2.29 (a), the intersection of plane P and plane V is indicated as P_v, frontal piercing line (正面迹线); the intersection of plane P and plane H is indicated as P_H, horizontal piercing line (水平迹线); the intersection of plane P and plane W is indicated as P_W, profile piercing line (侧面迹线). Fig. 2.29 (b) shows piercing lines, P_v, P_H and P_W in an unfold three projection planes system. The area between piercing line and two axes is the projection of the plane. For example, the area surrounded by P_v, axes OX and OZ is the front view of the plane P. With the same reason, the top view of the plane P is the area surrounded by P_H, axes OX and OY_H; the left view of the plane P is the area surrounded by P_W, axes OZ and OY_W. The point on the piercing line does not only exist in the plane P, but also in a projection plane.

As shown in Fig. 2.29 (a), line NM is in plane P. If n' as the front view of the endpoint N and m as the top view of the endpoint M are given, make the three views of line NM (Fig. 2.29(b)).

Analysis is as follows.

$\because n' \in P_v$, $\therefore$ endpoint $N \in$ plane P and plane V, $\therefore Y_N = 0$, $\therefore n \in$ axis OX, $n'' \in$ axis OZ. Similarly, $\because m \in P_v$, $\therefore$ endpoint $M \in$ plane P and plane H, $\therefore Z_M = 0$, $\therefore m' \in$ axis OX and $m'' \in$ axis OY_W.

Graphic solution is as follows (Fig. 2.29(c)).

Step 1　Through n', draw projection lines to intersect axis OX at n and axis OZ at n''.

Step 2　Throug m, draw projection lines to intersect axis OX at m' and axis OY_W at m''.

Step 3　Connect n' and m', n and m as well as n'' and m'' with thick lines to form the three views of line NM ($n'm'$, nm, $n''m''$).

Usually, piercing lines are used to represent a spatial plane located in particular position. As

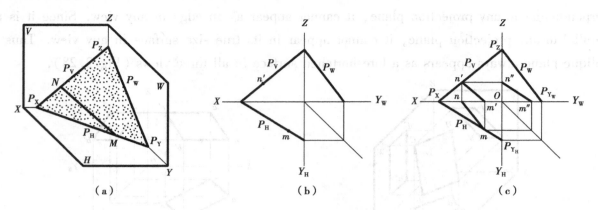

Fig. 2.29 Piercing lines of a plane.

discussed before, if a plane is perpendicular to a projection plane, its projection appears as an edge in the projection plane, so the piercing line that appears as an edge is able to depict (描述) a plane perpendicular to projection plane.

Figs. 2.30 (a) and (b) shows plane P is perpendicular to plane V and its three piercing lines are P_V, P_H, and P_W. Among P_V, P_H, and P_W, frontal piercing line P_V is one view that appears as an edge. Plane P is positioned as long as P_V is determined, so only P_V is used to express plane P as shown in Fig. 2.30 (c). Sometimes in order to draw simply, the piercing line appearing as an edge is broken into two short line segments shown as Fig. 2.30 (d).

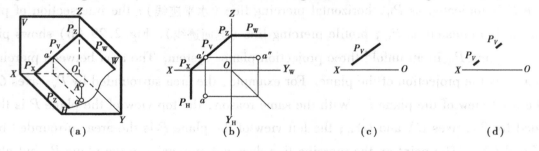

Fig. 2.30 Representation of a plane perpendicular to the plane V.

In Figs. 2.31 (a) and (b), plane P is parallel to plane H and it appears as an edge on the frontal piercing line and profile piercing line, respectively. Any one of the two piercing lines, P_V or P_W, can be used to represent plane P. Figs. 2.31 (c) and (d) show frontal piercing line P_V in the unfolded two projection planes system.

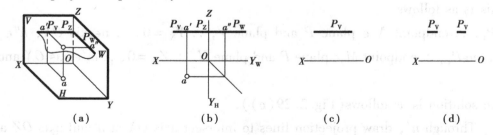

Fig. 2.31 Piercing line representation of a plane parallel to plane H.

2.5.4 Points and lines in a plane

According to the following axiom (公理) in geometry, points and lines in planes are picked up.

If a line is known to be in a plane, then any point on that line is in the plane. A line may be drawn in a plane by keeping it in contact with (intersecting) any two given lines in the plane. A line may also be located in a plane by drawing the line through a known point in the plane and parallel to a line in the plane.

In Figs. 2.32 (a) and (b), point K lies on line BC that is located in plane ABC, therefore point K is in plane ABC.

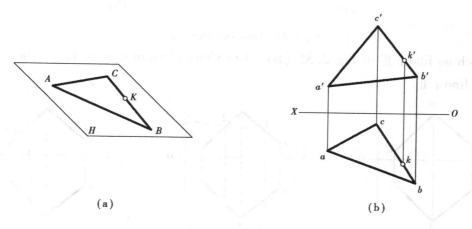

Fig. 2.32 Points in a plane.

In Figs. 2.33 (a) and (b), point N lies on line AB located in plane ABC, point M lies on line AC located in plane ABC, therefore line MN is located in plane ABC.

In Figs. 2.34 (a) and (b), plane Q is determined by two intersecting lines DE and EF. Point M is on line DE, line MN is drawn through M and parallel to the other line EF located in the plane, therefore line MN is located in plane Q.

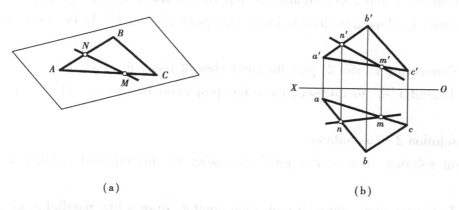

Fig. 2.33 Lines in a plane (I).

With above principle in geometry, a point may be placed in a plane by locating it on a line in the plane.

Example 2.7 Given e as the top view of point E in the plane determined by two intersecting lines AB and BC, find the front view of point E (Fig. 2.35 (a)).

Analysis is as follows.

An infinite number of lines containing point E exist in the plane. Any convenient one may be

45

Fundamentals of Engineering Drawing

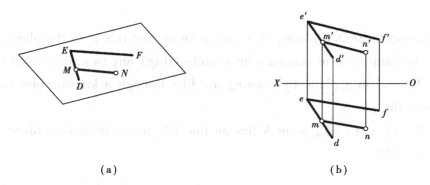

Fig. 2.34 Lines in a plane (II).

selected, such as line I II in Fig. 2.35 (b). The views of point E must lie on the corresponding views of the line I II.

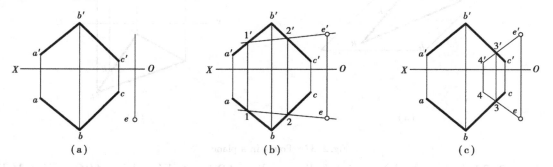

Fig. 2.35 Finding the front view of point E.

Graphic solution 1 is as follows.

Step 1　Through top projection point e, draw arbitrary line to intersect ab and bc at 1 and 2, respectively.

Step 2　Connect 1 and 2 to complete the top view of line I II, 1-2, (Fig. 2.35 (b)). From top projection point 1, draw a projection line to intersect $a'b'$ at $1'$. In the same way, $2'$ can be obtained.

Step 3　Connecting $1'$ and $2'$ gets the front view of line I II, $1'$-$2'$.

Step 4　Extend $1'$-$2'$ to intersect the other projection line from e at e', the front view of point E.

Graphic solution 2 is as follows.

A different solution of a similar problem—using the principle of parallelism is shown in Fig. 2.35(c).

Step 1　In the top view, through projection point e, draw a line parallel to ab to intersect bc at 3, that is, $4e \mathbin{/\mkern-2mu/} ab$.

Step 2　The top projection point 3 is projected to the front view to obtain $3'$.

Step 3　Through projection point $3'$, draw line $4'e'$ parallel $a'b'$, that is, $4'e' \mathbin{/\mkern-2mu/} a'b'$. Note that the length of $4'e'$ and $4e$ is arbitrary, but it is necessary that projection line is perpendicular to projection axis, that is $4'4 \perp OX$, $e'e \perp OX$.

Example 2.8　Given m' as the front view of point M in plane $ABCD$ perpendicular to plane H, find the top view of point M (Fig. 2.36 (a)).

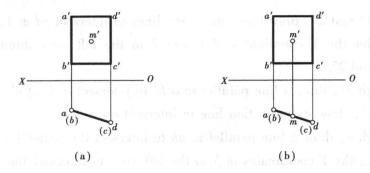

Fig. 2.36 Finding the top view of point M.

Analysis is as follows.

In Fig. 2.36 (a), the given two views show that plane ABCD is perpendicular to plane H. The plane ABCD appears as an edge in the top view, so the top view of the point M must be on the edge.

Graphic solution is as follows.

From m', draw its projection line to intersect the edge $a(b)(c)d$ at m, the top view of point M (Fig. 2.36(b)).

Similarly, in the Fig. 2.30 (b), if a is given as the top view of the point A in plane P which is perpendicular to plane V, the front view and left view of the point A may be worked out. Through a, draw projection line to intersect P_V at a' first and then transfer Y coordinates of the a to the left view through 45°miter-line to intersect the other projection line from a' at a''. a' and a'' are the front and left views required.

In Fig. 2.31 (b), if a is given as the top view of point A in plane P which is parallel to plane H, other two views of the point A can be completed. Through a, draw projection lines to intersect P_V and P_W at a' and a'', respectively. a' and a'' are the front and left views required.

Example 2.9 Complete the top and front views of figure ABCD with a gap (Fig. 2.37 (a)).

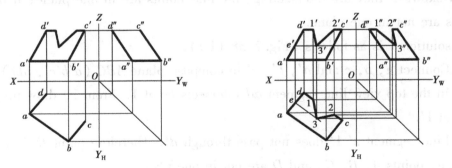

Fig. 2.37 Completing the top and front views of figure ABCD.

Analysis is as follows.

In F2.37(a), the given views show that figure ABCD is an oblique plane. In Fig. 2.37(b), the gap is assigned by $1'$, $2'$ and $3'$, in the front view. The top views 1 and 2 are obtained directly by means of $1'$ and $2'$ according to the projection characters of points on a line. But the top view 3 is obtained indirectly by $3'$ since $3'$ is located in area $a'b'c'd'$.

Graphic solution is as follows.

Step 1 From 1' and 2', draw their projection lines to intersect cd at 1, 2.

Step 2 Transfer the Y coordinates of 1 and 2 to the left view through 45° miter-line to intersect $c''d''$ at 1" and 2".

Step 3 Through 3', draw a line parallel to $a'b'$ to intersect $a'd'$ at e'.

Step 4 From e', draw its projection line to intersect ad at e.

Step 5 Through e, draw a line parallel to ab to intersect the projection line from 3' at 3.

Step 6 Transfer the Y coordinates of 3 to the left view to intersect the other projection line from 3' at 3".

Step 7 Connect 1, 3 and 3, 2 with thick line to complete the top view of the gap. Connect 1", 3"and 3", 2" with thick line to complete the left view of the gap.

Example 2.10 The front and top views of four points are given, judge whether the four points lie in one plane or not (Fig. 2.38(a)).

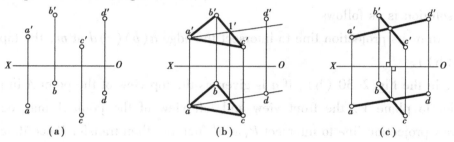

Fig. 2.38 Judging whether the four points lie in one plane or not.

Analysis is as follows.

Because three points are not on a straight line, it is able to construct a plane by connecting points A, B, and C, then judge whether the fourth point D is in this plane. Another approach is to construct two intersecting lines with four points, and observe whether the two lines are intersecting or skew. If they are intersecting, the four points are in one plane; if they are skew, the four points are not in one plane.

Graphic solution 1 is as follows (Fig. 2.38 (b)).

Step 1 Connect a, b, c and a', b', c' to compose plane ABC ($a'b'c'$, abc).

Step 2 In the top view line segment ad intersects bc at 1. From 1, draw projection line to intersect $b'c'$ at 1'.

Step 3 Line segment a' 1' does not pass through d'. Therefore point D does not lie in the plane ABC, i.e, points A, B, C, and D are not in one plane.

Graphic solution 2 is as follows (Fig. 2.38 (c)).

Connect a, d and a', d' to obtain line AD ($a'd'$, ad). Similarly, connect b, c and b', c' to obtain line BC ($b'c'$, bc). Obviously, two crossing points do not lie on a projection line, so lines AD and BC are skew. Therefore points A, B, C, and D are not in one plane.

2.5.5 Principal lines in a plane (投影面的平行线)

Principal lines in a plane include frontal lines, horizontal lines and profile lines. Many problems in practice require the addition of frontal, horizontal, or profile lines in a plane.

In Fig. 2.39 (a), the frontal line A is located in the plane ABC by first drawing its top view a1 parallel to axis OX, and then projecting 1 to b'c' to obtain 1', thus establishing a' 1'. Line A I (a'1', a1) is the required frontal line in the plane ABC.

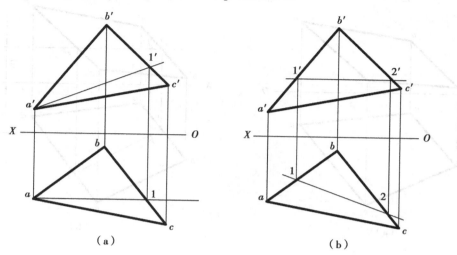

Fig. 2.39 Principal lines in a plane.

In a similar fashion, a horizontal line may be located by first drawing its front view in a horizontal position and then determining its top view by projection. For example, in Fig. 2.39 (b), at any height, draw a line parallel to axis OX to intersect a'b' and b'c' at 1', 2' respectively. From 1' draw projection line to intersect ab at 1. Similarly, from 2', may obtain 2 by projecting it to bc. Line I II (1'-2', 1-2) is the required horizontal line.

The application of principal lines in a plane is to locate a point in a plane.

Example 2.11 In a given plane ABCD, locate a point K which is 10 mm above plane H and 15 mm in front of plane V (Fig. 2.40(a)).

Analysis is as follows.

In the given plane, if construct a horizontal line which is 10 mm above plane H, this line is the locus (轨迹) of all points in the given plane which are 10 mm above plane H in space. Similarly, if establish a frontal line which is 15 mm in front of plane V, this line is the locus of all points in the given plane which are 15 mm in front of plane V in space. The intersecting point of the horizontal line and the frontal line is the required point K.

Graphic solution is as follows.

Step 1 A parallel line of axis OX is drawn where Z coordinates value equals 10 mm and it intersects a'b' and c'd' at e' and f', respectively.

Step 2 From e', draw its projection line to intersect ab at e. Through f' draw its projection line to intersect cd at f.

Step 3 Connecting e, f obtains the top view ef. Line EF (e'f', ef) is the required horizontal line in the given plane (Fig. 2.40(b)).

Step 4 A parallel line of axis OX is drawn where Y coordinates value equals 15 mm and it

Fundamentals of Engineering Drawing

intersects *ef* at *k*.

Step 5　From *k*, draw its projection line to intersect *e'f'* at *k'*.

Point *K* (*k'*, *k*) is the point required.

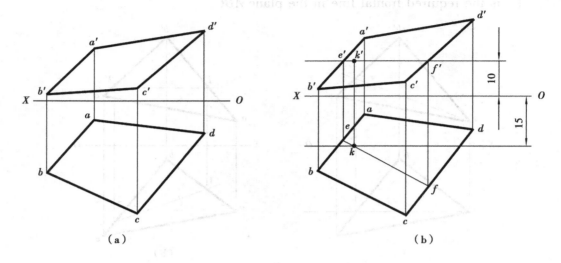

Fig. 2.40　A point in a plane.

Chapter 3　Primary Objects

Primary objects include polyhedra (平面立体) and revolutions (回转体). A composite object (组合体) can be regarded as a combination of primary objects. Thus, students should be familiar with projections as well as their drawing methods of those relative simple objects before being involved in more complex objects. This chapter will introduce the methods of drawing projections (views) of those objects.

3.1　Polyhedra (多面体)

A polyhedron is a multisided solid formed by intersecting planes. Planar surfaces that enclose a polyhedron are called lateral surfaces (侧棱面) or surfaces simply. Lines of intersection of adjacent surfaces are called edges (棱). To draw the views of a polyhedron is equivalent to draw the views of these surfaces and edges.

Meanwhile, it is necessary to judge the visibility of those elements. Usually, visible edge is drawn by thick line and invisible edge in hidden line. The method to find points or lines on the surface of a polyhedron is the same as that to find points or lines on a plane as stated in 2.5.4.

Typical polyhedra are prisms (棱柱体) and pyramids (棱椎体). In this book, the prisms and pyramids are right prisms (正棱柱) and right pyramids (正椎体), which are simply called prisms and pyramids.

3.1.1　Prisms (棱柱体)

A prism is a solid with two equal parallel polygon bases and several rectangular lateral surfaces. All edges in prism are perpendicular to bases. Different prisms have different polygon bases. For example, a triangular prism (三棱柱) has a triangular bases (Fig. 3.1); a rectangular prism (四棱柱) has rectangular bases (Fig. 3.2 (a)); and so on. Pentagonal prism (五棱柱) and hexagonal prism (六棱柱) are shown in Figs. 3.2 (b) and (c).

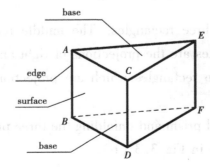

Fig. 3.1　Triangular prism.

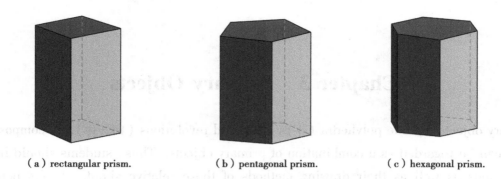

(a) rectangular prism.   (b) pentagonal prism.   (c) hexagonal prism.

Fig. 3.2 Prisms.

3.1.1.1 Three views of a prism

Take a hexagonal prism as an example to illustrate the way of drawing three views of a prism (Figs. 3.3 (a) to (c)).

Suppose a hexagonal prism is put properly, where its bases are parallel to plane H while its front and rear surfaces are parallel to plane V (Fig. 3.3(a)), then the object is projected toward three projection planes to generate three views: front view, top view and left view.

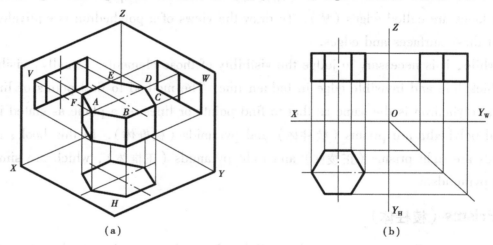

Fig. 3.3 Three views of a hexagonal prism.

1) Analyzing three views

The top view is a hexagon (六边形) that represents the true shape of two bases. In the hexagon, six sides are convergent lines of six lateral surfaces while the six corners are convergent points of six edges.

The front view consists of three rectangles. The middle rectangle is the projection of the middle surface. The rest rectangles are the projections of right and left surfaces.

The left view consists of two rectangles which are projections of the front-left and rear-left surfaces.

After removing the hexagonal prism and unfolding the three projection planes, the three views of the hexagonal prism are shown in Fig. 3.3 (b).

2) Steps of drawing three views

Step1 Draw projection axes with thin lines and symmetric centerlines with center lines as shown in Fig. 3.4(a).

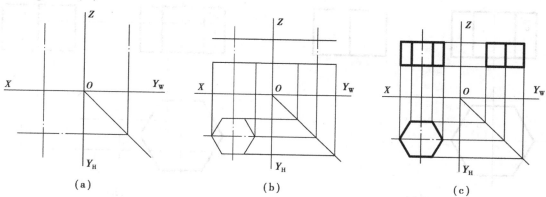

Fig. 3.4 Steps for drawing.

Step 2 Draw three views of the top and bottom bases (Fig. 3.4(b)).

Step 3 Draw three views of edges (Fig. 3.4(c)).

Step 4 Check, clean-up unnecessary lines and brighten the all visible lines with proper pencil such as B pencil or 2B pencil.

3) Judging visibility of surfaces

In the front view, visible surfaces are the hexagonal prism's front surface, front-left surface and front-right surface while rear, rear-left and rear-right surfaces are invisible. In the top view, visible surface is the hexagonal prism's top base while bottom base is invisible. In the left view, visible surfaces are the hexagonal prism's front-left and rear-left surfaces while front-right and rear-right surfaces are invisible.

Note that projection axes may not be drawn in engineering drawings, however, they exist actually. Namely, engineers default (默认) projection axes existing in adjacent views without drawing them.

3.1.1.2 Points and lines on the surfaces of a prism

The problem refers that when one view of a point or a line on certain lateral surface of a prism is given, complete other views of the point or the line.

The way to solve the problem is as follows.

Firstly, judge the spatial position of the point according to the visibility of the given projection point. If the given projection point is visible, the spatial point must be in the visible surface. If the given projection point is invisible, the spatial point must be in the invisible surface.

Thirdly, judge the visibility of the solution. If the surface or the edge with the solution is visible, the solution is visible. If the surface or the edge with the solution is invisible, the solution is invisible.

Besides, it is unnecessary to judge the visibility of the solution on the convergent line of a surface except the coincident points.

Example 3.1 Given m' and (n), as the front view of point M and the top view of point N as shown in Fig. 3.5(a), complete other views of points M and N.

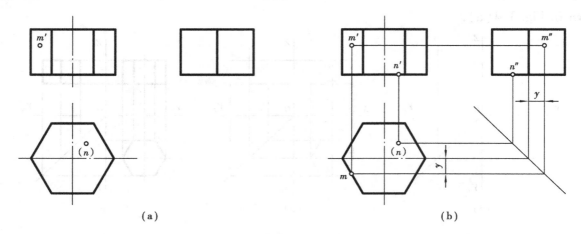

Fig. 3.5 Other views of points M and N.

Analysis is as follows.

In Fig. 3.5(a), the given three views show that the object is a hexagonal prism of which six lateral surfaces are perpendicular to plane H. Because m' is visible, point M is on the front-left surface of the hexagonal prism. Since (n) is invisible, point N should be on the bottom base.

Graphic solution is as follows.

Step 1 From m', draw projection line to top view to intersect the convergent line of front-left surface of the hexagonal prism at m.

Step 2 Transfer y—the Y coordinates of m, from the top view to left view through 45° miter-line to intersect the other projection line from m' at m''. The left view m'' is visible because the left view of the front-left surface is visible.

Up to now, m and m'' as other views of point M are completed.

By the same way, from (n), draw projection line to the front and left views to intersect the convergent line of bottom base at n' and n'', respectively.

Example 3.2 Given 1' 2' and 2' 3' as the front views of the lines I II and II III on the triangular prism's surfaces, complete other views of the lines (Fig. 3.6(a)).

Analysis is as follows.

In Fig. 3.6(a), the given three views show that the object is a triangular prism of which two bases are parallel to plane W and three lateral surfaces are perpendicular to plane W. Thus, three lateral surfaces appear as edges in the left view to form a triangle and the left views of lines I II and II III are located on the sides of the triangle, Only the top views of lines I II and II III need to be completed.

The given three views also show that point I belongs to the top edge of the triangular prism because 1' is on the front view of the top edge. Similarly, point II belongs to the middle edge of the triangular prism. However, point III is located in the lower lateral surface of the triangular prism because 3' is located in an area which is the front view of the lower lateral surface.

Graphic solution is as follows.

Step1 Through 1' and 2', draw projection lines to the top view to intersect corresponding edges at 1 and 2. Evidently, horizontal projection point 1 and 2 are visible.

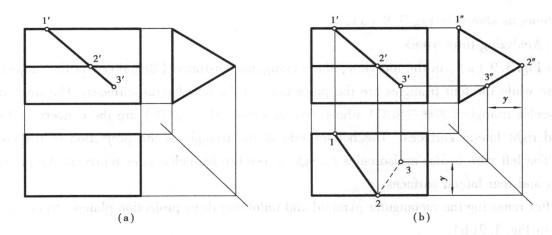

Fig. 3.6 Complete other views of the lines.

Step 2 From 3′, draw the projection line to the left view to intersect the oblique line at 3″.

Step 3 Transfer y—the Y coordinates of the 3″, to the top view to intersect the other projection line drawn from 3′ at (3). Because the top view of the lower lateral surface is invisible, the top view of the point Ⅲ is invisible and indicated by (3).

Step 4 Connect points to form lines. Connect 1, 2 with a thick line and connect 2, (3) with a hidden line to obtain the top view of the lines.

Finished drawing is shown in Fig. 3.6 (b).

3.1.2 Pyramids (棱锥体)

A pyramid is a solid with a polygon base and several triangular lateral surfaces that converge at a point, vertex. Different pyramids have different polygon bases. For example, a triangular pyramid (三棱锥) has a triangular base (Fig. 3.7). Similarly, a square pyramid (四棱锥) has square base (Fig. 3.8 (a)). A pentagonal pyramid (五棱锥) is shown in Fig. 3.8 (b).

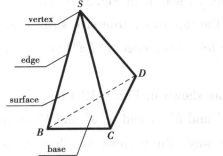

Fig. 3.7 Triangular pyramid.

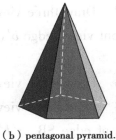

(a) rectangular pyramid.　　(b) pentagonal pyramid.

Fig. 3.8 Pyramids.

3.1.2.1 Three views of a pyramid

A rectangular pyramid is taken as an example to illustrate the method of drawing the three views of pyramid.

Suppose a rectangular pyramid is put properly, where its bases are parallel to plane H while its left and right lateral surfaces are perpendicular to plane V; its front and rear surfaces are perpendicular to plane W, then the object is projected toward three projection planes to generate three views as shown in Fig. 3.9 (a).

1) Analyzing three views

In Fig. 3.9 (a), in the top view, the rectangular out-frame (矩形外框) is the projection of the base while the four triangles are the projections of the four lateral surfaces. The front view is an isosceles triangle (等腰三角形) whose two isosceles sides (两腰) are the convergent lines of left and right lateral surfaces. The bottom side of the triangle is the projection of the object's base. The left view is also an isosceles triangle whose two isosceles sides represent the projections of front and rear lateral surfaces.

After removing the rectangular pyramid and unfolding three projection planes, three views are shown in Fig. 3.9(b).

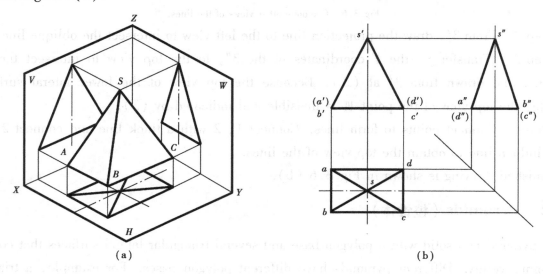

Fig. 3.9 Three views of a rectangular pyramid.

2) Steps of drawing three views

Step 1   Draw symmetric centerline with center lines as shown in Fig. 3.10 (a).

Step 2   Draw three views of the base. First draw the top view, true size surface abcd, and then the front view, edge $b'(a')(d')c'$, as well as the left view, edge $a''(d'')(c'')b''$ (Fig. 3.10 (b)).

Step 3   Draw three views of the vertex, s', s, s'' as shown in Fig. 3.10 (b).

Step 4   Draw three views of the edges. Connect s' and b', s and b, s'' and b'' to obtain the three views of edge SB, $s'b'$, $s b$, $s''b''$. By the same way, three views of other edges can be obtained. Meanwhile, three views of four lateral surfaces can be obtained.

Lastly, check, clean-up unnecessary and brighten all the visible lines to complete three views.

3) Judging visibility of surfaces

In the front view, visible surface is the rectangular pyramid's front surface while the rear surface is invisible. In the top view, visible surfaces are the rectangular pyramid's four surfaces while the bottom base is invisible. In the left view, visible surface is the rectangular pyramid's left surface while right surface is invisible.

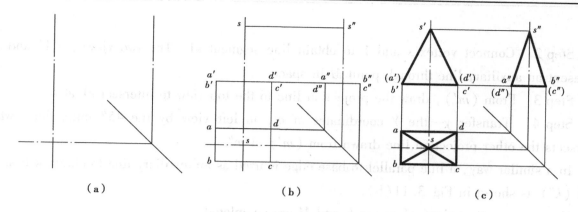

Fig. 3.10 Steps for drawing.

3.1.2.2 Points and lines on the surface of pyramid

The problem and the way to solve the problem are similar to those of a prism. The difference between them is that there may be some oblique surfaces on the pyramid. When a point is on an oblique surface, through the point, draw an auxiliary line first and then complete the view of the point.

Example 3.3 Given (m'), k' as the front views of points M and K as shown in Fig. 3.11 (a), complete other views of the points.

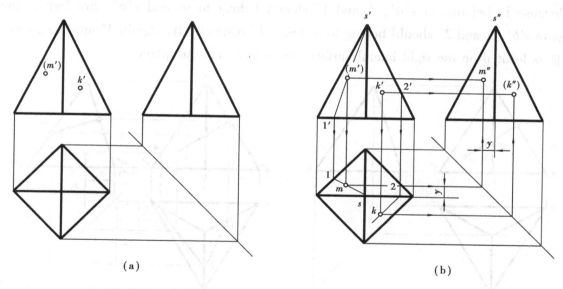

Fig. 3.11 Other projections of points K and M.

Analysis is as follows.

In Fig. 3.11(a), the given three views show that the object is a rectangular pyramid of which the base is parallel to plane H while four surfaces are oblique planes. As (m') appears as invisible, point M is located on the rear-left surface which is an oblique plane. As k' is visible, point K is located on the front-right surface which is also an oblique plane too.

Graphic solution is as follows.

Three views of the vertex are s', s, s''.

Step 1 In front view, connect vertex s' and (m') to obtain line segment $s'(m')$ and extend it to intersect base at $(1')$. Through $(1')$, draw projection line to the top view to intersect base at 1.

Step 2   Connect vertex s and 1 to obtain line segment s1. The two views, s'1' and s1, represent an auxiliary line through point M in space.

Step 3   From (m'), draw the projection line to the top view to intersect s1 at m.

Step 4   Transfer y—the Y coordinates of m, to left view by the 45° miter-line, which intersects the other projection line drawn from (m') at m".

In a similar way, a line parallel to base edge is used as an auxiliary line to obtain k first and then (k") as shown in Fig. 3.11(b).

Up to now, other views of points K and M are completed.

Example 3.4   In Fig. 3.12(a), given 1' 2' and 2' 3' as the front views of lines I II and II III, complete the other views of the lines.

Analysis is as follows.

As shown in Fig. 3.12 (a), the given three views show that the triangular pyramid's base is parallel to plane H while left and right lateral surfaces are oblique planes. The rear lateral surface is inclined plane perpendicular to plane W since in the left view, edge AC converge to a point a"(c").

Because 1' belongs to s'a', 1 and 1" should belong to sa and s"a". Similarly, since 2' belongs to s'b', 2 and 2" should belong to sb and s"b" respectively. Again 3' appears as visible, point III is located on the right lateral surface which is an oblique plane.

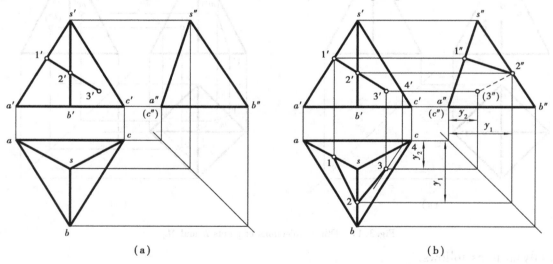

Fig. 3.12   Other views of the line I II and line II III.

Graphic solution is as follows.

Step1   From 1', draw its projection line to the top and left views respectively, which intersect sa at 1 and s"a" at 1".

Step 2   Through 2', draw its projection line to the left view to intersect s"b" at 2" first and then transfer y_1—the Y coordinates of 2", to the top view to intersect sb at 2.

Up to now, the three views of the point I (1', 1, 1"), point II (2', 2, 2") are completed.

How to find out the top and left views of point III? Because point III is on the oblique plane, it is necessary to make an auxiliary line to complete its top view.

Step 3　From 3′, draw an auxiliary line parallel to $b'c'$ to intersect $s'c'$ at 4′.

Step 4　From 4′, draw its projection line to the top view to intersect sc at 4.

Step 5　Through 4, draw a line parallel to bc to intersect the projection line drawn from 3′ at 3.

Step 6　Transfer y_2—the Y coordinates of 3, to the left view to intersect the other projection line drawn from 3′ at (3″). Note that (3″) is invisible because area $s''(c'')b''$ is invisible.

Step 7　Connect points to form lines. In the top view, connect 1-2 and 2-3 with thick line. In left view, connect 1″2″ in thick line and connect 2″(3″) with hidden line.

Lastly, check, clean-up the unnecessary lines and brighten the visible lines. Finished drawings are shown in Fig. 3.12(b).

3.2　Revolutions（回转体）

Revolutions such as cylinders（圆柱）, cones（圆锥）, spheres and toruses（圆环）are typical primary objects in engineering.

A revolution is generally composed of a revolutionary surface which is generated by a moving line (straight line or curve) revolving around a fixed line called axis（轴线）(Fig. 3.13). The moving line is called generatrix（母线）. Each position of the generatrix is called an element（素线）. A latitude-circle（纬圆）is a circle, which is locus（轨迹）of an arbitrary point on the generatrix, always perpendicular to the revolution' axis. For example, in Fig. 3.13, generatrix SA sweeps around axis OO to generate a revolutionary surface, conical surface. The locus of an arbitrary point K on generatrix SA forms a latitude-circle. SA_1 is an element-line.

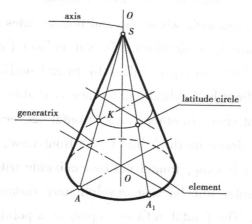

Fig. 3.13　Formation of a conical surface.

3.2.1　Cylinders

A cylinder is a solid with two circular plane bases and a revolutionary surface called cylindrical surface generated by a straight line (generatrix) sweeping around an axis with constant distance (Fig. 3.14 (a)). Each position of the generatrix is called an element of the cylinder.

All elements on the cylindrical surface are parallel to the axis.

3.2.1.1 The three views of a cylinder

If the cylinder is put properly, where its axis is perpendicular to one of the three projection planes such as plane H, as a result, two circular bases will be parallel to the plane H as shown in Fig. 3.14(b). The cylinder is projected toward three projection planes to form three views: a circular view and two rectangular views.

1) Analyzing the three views and their visibility

The cylinder's top view is a circle showing the true sizes of the upper and bottom bases. The upper base is visible while bottom base is invisible. Cylindrical surface appears as a circumference in the top view.

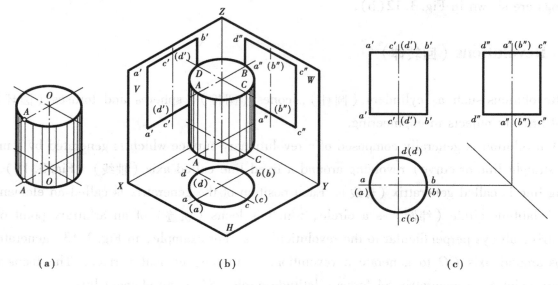

Fig. 3.14 Three views of a cylinder.

Cylinder's front view is a rectangle whose top and bottom sides are the projections of circular bases and the other two sides are the projections of frontal outlines (正面外形线) AA and BB. The cylindrical surface is divided into two equal halves by frontal outlines symmetrically, which are front half and rear half. The front half cylindrical surface is visible while the rear half cylindrical surface is invisible in the front view. In other words, frontal outlines are located on the front-rear symmetric plane and they are drawn by thick line in the front view, e.g. $a'a'$ and $b'b'$ as shown in Fig. 3.14(b). But in the left view, frontal outlines coincide with the centerline and they are substituted (替代) by the center line because revolutionary surface is smooth, e.g. $a''a''$ and $(b'')(b'')$ in Fig. 3.14 (b). The frontal outlines appear as a point respectively in the top view, e.g. $a(a)$ and $b(b)$.

The left view of the cylinder has a similar situation with the front view. Profile outlines (侧面外形线) CC and DD divide cylindrical surface into left half and right half. The former is visible while the later is invisible in the left view. In other words, profile outlines are located on the left-right symmetric plane and they are drawn by thick line in the left view, e.g. $c''c''$ and $d''d''$ as shown in Fig. 3.14(b). But in the front view, they coincide with the centerline and they are substituted by the centerline, e. g. $c'c'$ and $(d')(d')$ in Fig. 3.14(b) while in the top view, they appear as a point respectively, e. g. $c(c)$ and $d(d)$.

After removing the cylinder and unfolding three projecting planes, three views are shown in Fig. 3.14 (c).

2) Steps of drawing the three views

Step 1   Draw the three views of the axis of the cylinder with centerlines. Draw symmetric centerline in the top view with center lines (Fig. 3.15(a)).

Step 2   Draw the three views of the upper and lower circular bases as shown in Fig. 3.15 (b).

Step 3   Draw the three views of frontal and profile outlines.

Lastly, check, clean-up the unnecessary lines and brighten all the visible lines.

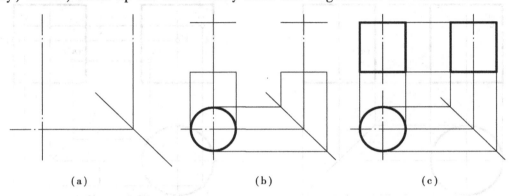

Fig. 3.15   Steps for drawing.

3.2.1.2   Points and lines on a cylinder

Given one view of a point on the surface of a cylinder, complete the other views of the point?

Firstly, judge the spatial position of the point according to the visibility of the given projection point. Suppose that the axis of the cylinder is perpendicular to plane H. When the given projection point is in the front view of the cylinder and it is visible, the spatial point is on the front half of the cylinder; but if it is invisible, the spatial point is on the rear half of the cylinder. When the given projection point is in the left view of the cylinder and it is visible, the spatial point is on the left half of the cylinder; but if it is invisible, the spatial point is on the right half of the cylinder. When the given projection point is in the top view of the cylinder and it is visible, the spatial point is on the upper base of the cylinder; but if it is invisible, the spatial point is on the bottom base of the cylinder. When the given projection point is on the outline of the cylinder, the spatial point is on the symmetric plane of the cylinder, and vice versa (反之亦然).

Secondly, complete the other views. If the spatial point is on the outline of the cylinder, its other two views can be obtained at the same time. If the spatial point is anywhere on the cylindrical surface other than on the outline, the view on the circumference is found out first and then the other view.

Thirdly, it is required to judge the visibility of the solution. If the surface with the solution is visible, the solution is visible. If the surface with the solution is invisible, the solution is invisible.

Besides, it is unnecessary to judge the visibility of the solution on the circumference except coincident points.

Example 3.5 Given m', as the front view of point M; (n'') and k'', as the left views of points N and K, complete other views of points (Fig. 3.16(a)).

Analysis is as follows.

In Fig. 3.16(a), the given three views show that the object is a cylinder whose axis is perpendicular to plane H. From m', (n'') and k', it is known that the point M locates on the front-left cylindrical surface; point N on the rear-right cylindrical surface; point K on the profile outline.

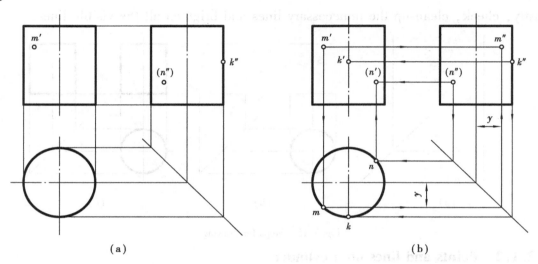

Fig. 3.16 Other views of points M, N, K.

Graphic solution is as follows.

First, m and m'' are found by m'.

Step 1 From m', draw its projection line to the top view to intersect the circumference at m, the top view of the point M.

Step 2 Transfer y—the Y coordinates of m, to the left view to intersect the other projection line drawn from m' at m'', the left view of the point M.

Second, k and k' are found by k''.

Through k'', draw its projection lines to the front and top views to intersect the center line which is the left-right symmetric plane, in space at k' and k respectively.

Third, n and (n') are gotten from n''.

Step 1 Through (n''), draw its projection line to the top view to intersect the circumference at n, the top view of the point N.

Step 2 Through (n'') draw its projection line to the front view to intersect the other projection line from n at n', the front view of the point N.

Up to now, the three points' other views are completed. See Fig. 3.16(b).

Example 3.6 Given $a' b' c'$ as the front view of curve ABC on the surface of the cylinder, complete other views of the curve (Fig. 3.17(a)).

Analysis is as follows.

In Fig. 3.17(a), the given three views show that the object is a cylinder whose axis is perpendicular to plane H. From front projection $a'b'c'$, it is known that the given curve is located on the front half of the cylinder. The front view also shows that point A is on the left frontal outline (左正面外形线), point B on the front profile outline (前侧面外形线) and point C on the front-right cylindrical surface. The curve ABC appears as an arc on the circumference in the top view. Thus, only the left view of the curve ABC needs to be found out.

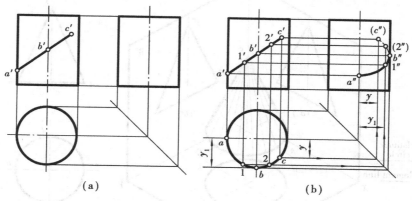

Fig. 3.17 Other views of curve ABC.

Graphic solution is as follows.

Step 1 Through a', draw its projection line to the left view to intersect the center line, front-rear symmetric plane, at a''.

Step 2 From b', draw its projection line to the left view to intersect the front profile outline at b''.

Step 3 Through c', draw its projection line to the top view to intersect the circumference at c first and then transfer y—Y coordinates of c, to the left view to intersect the other projection line from c' at (c'').

Step 4 Pick $1'$ properly on the line segment $a'b'c'$ and through it, draw its projection line to the top view to intersect the circumference at 1. Transfer the y_1—Y coordinates of 1, to the left view to intersect the other projection line from $1'$ at $1''$.

In the same way, the left views of a series of points can be obtained. For example, $(2'')$ are obtained by $2'$, which $(2'')$ is invisible, because from $2'$, it is known that the point II is located on the right half of the cylinder.

Step 5 Connect points to form a curve smoothly. Connect a'', $1''$, b'' with a thick line and connect b'', $(2'')$ and (c'') with a hidden line to form the left view of the curve, on which the b'' is the critical point (临界点) of visibility and invisibility (Fig. 3.17(b)).

3.2.2 Cones

A cone is a solid with a circular plane base and a revolutionary surface called conical surface generated by a straight line (generatrix) sweeping around the axis (Fig. 3.18(a)). If the straight line intersects the axis at an angle less than 90°, its locus is a conical surface and the intersection point is called the vertex of the cone. All elements on the conical surface converge to the vertex. All latitude-circles are always perpendicular to the axis of the cone. For example, if the axis is perpendicular to plane H, latitude-circles are a series of horizontal latitude-circles. Besides, the latitude-circle which is close to the vertex is smaller than that far from the vertex. At the vertex, the latitude circle becomes a point.

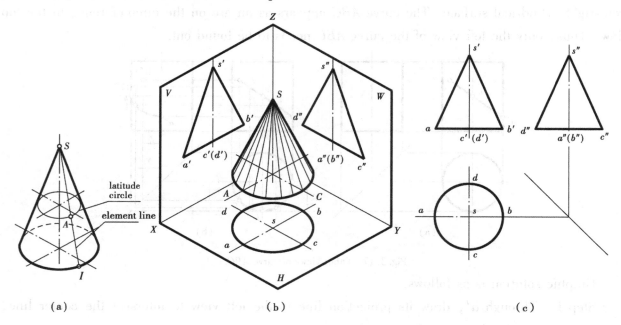

Fig. 3.18 Three views of a cone.

3.2.2.1 The three views of a cone

If a cone is put properly, where its axis is perpendicular to one of the three projection planes such as plane H, a circular base will be parallel to the plane H (Fig. 3.18(b)). And then the cone is projected toward three projection planes to form the three views: one circular view and two triangular views.

1) Analyzing the three views and their visibility

The cone's top view is a circle area which is the projections of the circular base and the conical surface. The conical surface is visible while the circular base is invisible.

The cone's front view is an isosceles triangle whose bottom side is the projection of the circular base and its length is equal to the diameter of the circular base. Two isosceles sides are the projections of frontal outlines SA and SB which divide the conical surface into two halves, the front half and the rear half. The front half is visible while the rear half is invisible. In other words, frontal outlines are located on the front-rear symmetric plane and they are drawn by thick lines in the front view, e.g. $s'a'$, $s'b'$ in Fig. 3.18(b). But in the left and top views, frontal outlines coincide with the centerline and they are substituted by the center line because of the smoothness of the revolutionary surface, e.g. sa, sb and $s''a''$, $s''(b'')$ in Fig. 3.18(b).

The cone's left view is similar to the front view. Profile outlines SC and SD divide the conical surface into left and right halves. The former is visible while the latter is invisible. In other words, profile outlines are located on the left-right symmetric plane and they are drawn by thick lines in the left view, e. g. $s''c''$, $s''d''$ in Fig. 3.18(b). But in the front and top views, they coincide with the centerline and they are substituted by the centerline, e. g. sc, sd and $s'c'$, $s'(d')$ in Fig. 3.18(b).

The three views are shown in Fig. 3.18(c) after removing the cone and unfolding the three projection planes.

2) Steps of drawing three views

Step 1 Draw the axis and symmetric centerline with centerlines as shown in Fig. 3.19(a).

Step 2 Draw the three views of the circular base as shown in Fig. 3.19(b).

Step 3 Draw frontal and profile outlines.

Step 4 Check, clean-up the unnecessary lines and brighten all the visible lines.

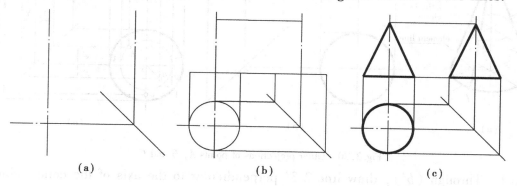

Fig. 3.19 Steps for drawing.

3.2.2.2 Points and lines on the surface of a cone

The problem and the way to solve the problem are similar to those of a cylinder. The difference between them is that the projection of a conical surface is not a circumference but an area. Thus, a spatial point is anywhere on the conical surface other than on the outline. Through the point, draw a element or a litiude circle first and then complete the view of the point.

Example 3.7 Given a' and (b') as the front views points A, B; c'', as the left view of point C, complete other views of points (Fig. 3.20(b)).

Analysis is as follows.

The given three views in Fig. 3.20(b) show that the object is a cone whose axis is perpendicular to plane H. From a', (b') and c'', it is known that the point A is located on the front-left conical surface, point B on the rear-right conical surface and point C on the profile outline.

Graphic solution is as follows.

The three views of the cone's vertex are s', s, s''.

An element is applied to find other projections of point A.

Step 1 Connect s', a' to obtain line segment $s'a'$, and extend it to intersect the bottom side of the cone at $1'$.

Step 2 From $1'$, draw its projection line to the top view to intersect the bottom circle at 1.

Step 3 Connect projection points s, 1. Line $S\text{ I}$ ($s1$, $s'1'$) is an element-line on the conical surface, which passes through point A (a, a').

Step 4 Through a', draw its projection line to the top view to intersect $s1$ at a, the top view of point A.

Step 5 Transfer y—the Y coordinates of a, to the left view to intersect the other projection line drawn from a' at a'', the left view of point A.

A latitude-circle can also be used to find other projections of point B.

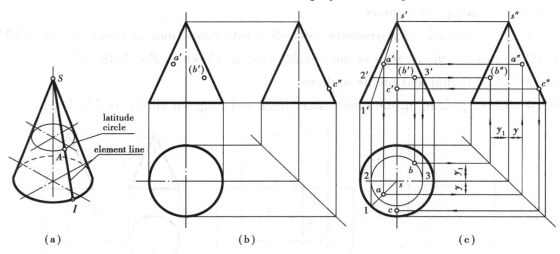

Fig. 3.20 Other projections of points A, B and C.

Step 1 Through (b'), draw line $2'3'$ perpendicular to the axis of the cone, which is the front view of the latitude-circle. The distance between $2'$ and $3'$ is the diameter of the latitude-circle.

Step 2 With vertex s as the center and line segment $2'3'$ as diameter, construct a circle in the top view.

Step 3 From (b'), draw its projection line to top view to intersect the latitude-circle at b, the top view of point B.

Step 4 Transfer y_1—the Y_1 coordinates of b, to the left view to intersect the other projection line drawn from (b') at (b''), the left view of Point B.

From c'', c' and c may be gotten directly as point C is on the profile outline. Through c'', draw projection lines to the front view and top view to intersect the centerlines which is the left-right symmetric plane, in space at c' and c respectively.

Lastly, check, clean-up the unnecessary lines and brighten all the visible lines. Up to now, other views of points A, B and C are obtained. See Fig. 3.20(c).

3.2.3 Spheres

A sphere is a solid with a revolutionary surface called spherical surface which is generated by a generatrix circle (母线圆) sweeping around one of its diameters. This diameter becomes the axis of the sphere and the ends of the axis are the poles of the sphere (Fig. 3.21(a)). Any diameter in the generatrix circle may be considered as an axis of a sphere. Thus, there are frontal latitude-circles, horizontal latitude-circles or profile latitude-circles on the spherical surface. For example, if the diameter perpendicular to plane V is regarded as the sphere's axis, spherical surface has a series of frontal latitude-circles (Fig. 3.22(a)). If the diameter perpendicular to plane H is considered as the sphere's axis, the spherical surface has a series of horizontal latitude-circles (Fig. 3.22(b)). If the diameter perpendicular to plane W is considered as the sphere's axis, the spherical surface has a series of profile latitude-circles (Fig. 3.22(c)).

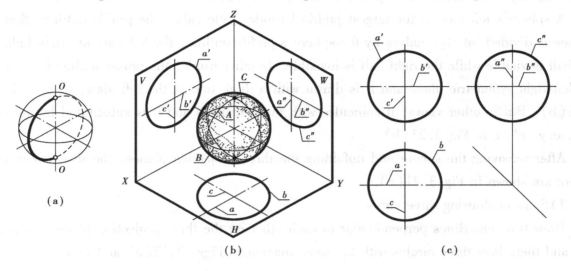

Fig. 3.21 Three views of a sphere.

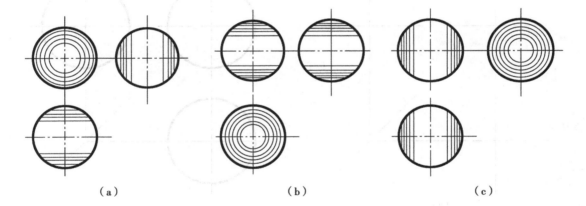

Fig. 3.22 Latitude-circles on a sphere.

3.2.3.1 The three views of a sphere

No matter which position a sphere is put at, its three views are three circles with the same diameter. See Fig. 3.21(b).

1) Analyzing the three views and their visibility

A sphere's front view is the largest frontal latitude-circle called the frontal outline. Spherical surface is divided into two halves by the sphere's frontal outline, the front half and the rear half. The front half is visible while the rear half is invisible. In other words, the frontal outline is located on the front-rear symmetric plane and it is drawn by a thick line in the front view, e.g. a' in Fig. 3.21(b). But in other views, it coincides with the center line and it is substituted by the center line because of the smooth ness of the revolutionary surface, e.g. a, a'' in Fig. 3.21 (b).

A sphere's top view is the largest horizontal latitude-circle called the horizontal outline. Spherical surface is divided into two halves by the sphere's horizontal outline, the upper half and the lower half. The upper half is visible while the lower half is invisible. In other words, the horizontal outline is located on the upper-lower symmetric plane and it is drawn by thick line in the top view e. g. b in Fig. 3.21(b). But in other views, it coincides with the centerline and it is substituted by the centerline, e. g. b', b'' in Fig. 3.21(b).

A sphere's left view is the largest profile latitude-circle called the profile outline. Spherical surface is divided into two halves by the sphere's profile outline, the left half and right half. The left half is visible while the right half is invisible. In other words, the profile outline is located on the left-right symmetric plane and it is drawn with a thick line in the left view, e. g. c'' in Fig. 3.21(b). But in other views, it coincides with the centerline and it is substituted by the center line, e. g. c', c in Fig. 3.21(b).

After removing the sphere and unfolding the three projection planes, the three views of the sphere are shown in Fig. 3.21(c).

2) Steps of drawing three views

Draw two centerlines perpendicular to each other on the three projection planes respectively first and then draw three circles with the same diameter (Figs. 3.23(a) and (b)).

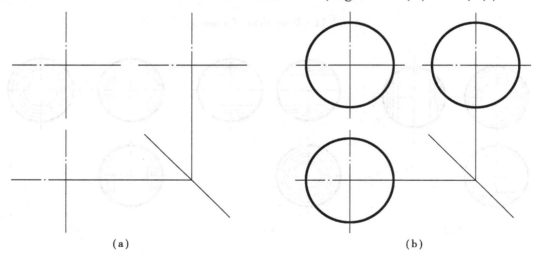

Fig. 3.23 Steps for drawing.

3.2.3.2 Points on a spherical surface

The problem and the way to solve the problem are similar to those of a cone (see 3.2.2.2). But their difference is that through a point on the spherical surface, some latitude-circles parallel to projection planes can be constructed, which may be front latitude-circle, top latitude-circle or profile latitude-circle.

Example 3.8 Given a as the top view of point A; b' as the front view of point B, complete other views of the points (Fig. 3.24(a)).

Analysis is as follows.

From b' and a, it is known that point B is located on the frontal outline and point A on the front-upper-right spherical surface.

Graphic solution is as follows.

The three views of the sphere's center are o', o, o''.

Because point B is on the frontal outline, other views may be gotten directly according to b'. Namely, from b', draw its projection lines to the top view and the left view to intersect the centerlines at (b) and b'', respectively (Fig. 3.24(b)).

To complete the front and left views of point A, the latitude-circle through point A must be constructed. Take the frontal latitude-circle as an example to illustrate.

Step 1 Draw a line perpendicular to the axis through a, which intersects the horizontal outline at 1 and 2. Line segment 12 is not only the top view of the frontal latitude-circle but also the diameter of the latitude-circle.

Step 2 With o' as center and segment line 12 as diameter, construct a frontal latitude-circle in the front view.

Step 3 Through a, draw its projection line to the front view to intersect the frontal latitude-circle at a', the front view of point A.

Step 4 Transfer y—the Y coordinates of a, to the left view to intersect the other projection line drawn from a' at (a''), the left view of point A (Fig. 3.24(b)).

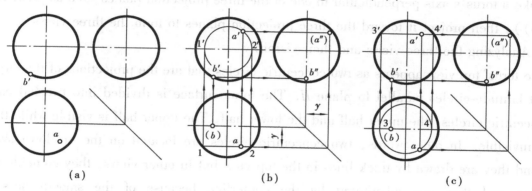

Fig. 3.24 Other projections of point A.

Fig. 3.24(c) shows how to get a' by means of the horizontal latitude-circle through a. The analysis of drawing procedure is left to the readers.

3.2.4 Toruses (圆环)

A torus is a solid with a revolutionary surface called torus surface which is generated by a generatrix circle sweeping around an axis that is eccentric (偏心的) to the generatrix circle (Fig. 3.25(a)). The distance between the center of generatrix circle and the axis is called eccentricity (偏心距). The generatrix circle at any position is called element-circle (素线圆). While sweeping of the generatrix circle, we can see that the locus of the centre of the generatrix circle is also a circle.

The interior torus surface (内环面) is formed by one half of generatrix circle close to the axis and the exterior torus surface (外环面) formed by the other half of the generatrix circle. As shown in Fig. 3.25 (a), the exterior torus surface is drawn by a thick line while the interior torus surface by a hidden line. In a torus, the latitude-circle is always perpendicular to the axis and there are two latitude-circles on the same level (同一层面上), a larger one at the exterior torus surface and a smaller one at the interior torus surface.

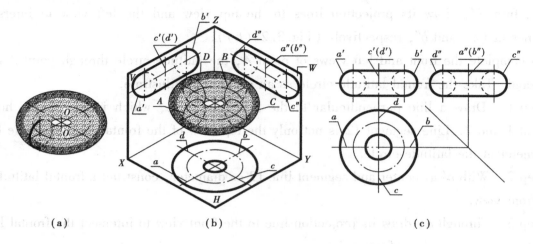

Fig. 3.25 Three views of a torus.

3.2.4.1 The three views of a torus

Make a torus's axis perpendicular to one of the three projection planes such as plane H (Fig. 3.25(b)), then project it toward the three projection planes to form the three views.

1) Analyzing the three views and their visibility

The torus' top view appears as two concentric circles that are the projections of the largest and smallest latitude-circles parallel to plane H. The torus surface is divided into two halves by the two concentric circles, the upper half and the lower half. The upper half is visible while the lower half is invisible. In other words, two concentric circles are located on the up-down symmetric plane and they are drawn by thick lines in the top view but in other views, they coincide with the centerline and they are substituted by the centerline because of the smooth ness of the revolutionary surface.

The torus' front view is composed of two small circles with equal diameter and two straight line segments tangent to the two small circles. The two small circles, a'' and b', are called frontal outlines which are the projections of two element-circles parallel to plane V, which are also the leftmost element-circle and the rightmost element-circle. The two straight lines are the projections of the highest latitude-circle and the lowest latitude-circle. The frontal outlines divided the torus into halves, the frontal half and the rear half. The exterior torus surface of the frontal half is visible while the exterior torus surfaces of the rear half and all the interior torus surface are invisible. The frontal outlines of the exterior torus surface are drawn with thick lines while the frontal outlines of the interior torus surface are drawn with hidden lines, e.g. a' and b' in Fig.3.25(c) but in other views, they coincide with the centerline and are substituted by the center line, e.g. a, b and a'' (b'') in Fig. 3.25 (c).

The torus' left view is similar to the front view. The two small circles, c'' and d'', are called profile outlines which are the projections of two element-circles parallel to plane W, which are also the forefront and rearmost element-circles. The two straight lines are the projections of the highest latitude-circle and the lowest latitude-circle. The profile outlines divide the torus into halves, the left half and the right half. In the left view, the exterior torus surface of the left half is visible while the exterior torus surfaces of the right half and all the interior torus surface are invisible. The profile outlines of the exterior torus surface are drawn with thick lines while the profile outlines of the interior torus surface are drawn with hidden lines, e. g. c'' and d'', in Fig. 3.25(c) but in other views, they coincide with the centerline and are substituted by the centerline, e. g. c, d and $c'(d')$ in Fig. 3.25 (c).

2) Steps of drawing the three views

Step 1   Draw the axis, symmetric centerline and locus circle of the center of the generatrix circle with center lines as shown in Fig. 3.26(a).

Step 2   Draw the three views of the largest latitude-circle and the smallest latitude-circle shown in Fig. 3.26(b).

Step 3   Draw outlines and two straight lines tangent to two small circles shown in Fig. 3.26 (b).

Step 4   Check, clean-up the unnecessary lines and brighten visible the lines.

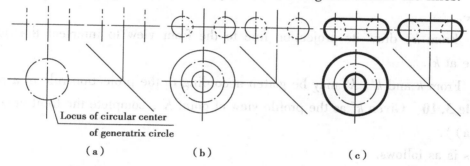

Fig. 3.26   Steps of drawing a torus.

3.2.4.2 Points on the torus surface

The problem and the solution to it are similar to those of a cone.

Example 3.9   Given k as the top view of point K, complete other views of the point (Fig. 3.27(a)).

Analysis is as follows.

In Fig. 3.27(a), the given three views show that the object is a quarter of the torus whose axis is perpendicular to plane V. The axis' top view coincides with a thick line which is the projection of the right side of the torus. From k, it is known that point K is located on the upper-rear exterior torus surface. Thus, it is necessary to make a latitude-circle through point K to find k'. The latitude-circle on the torus is parallel to plane V as the axis of the torus is perpendicular to plane V.

Graphic solution is as follows (Fig. 3.27(b)).

The three views of the torus' center are o', o, o''.

Step 1   In the top view, from k, draw a line perpendicular to the axis to intersect the horizontal outline at 1.

Step 2   Through 1, draw its projection line to the front view to intersect the centerline at $1'$.

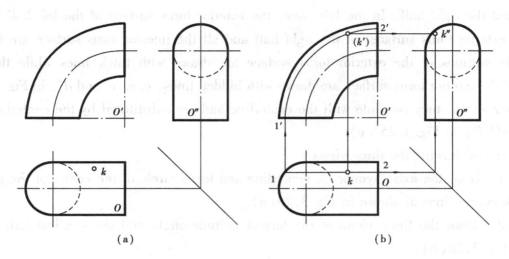

Fig. 3.27 Other views of point K.

Note that the center line coincides with a thick line partly.

Step 3 With o' as the center and line segment $o'1'$ as the radius, draw a quarter of a latitude-circle.

Step 4 From k, draw its projection line to the front view to intersect the quarter of the latitude-circle at k'.

Step 5 From k and k', k'' may be gotten according to the projection rule of a point.

Example 3.10 Given k'' as the profile view of point K, complete the front view of the point (Fig. 3.28(a)).

Analysis is as follows.

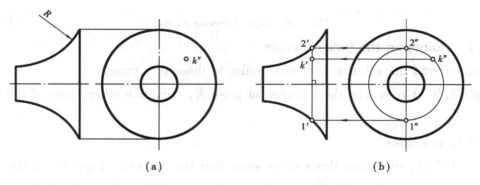

Fig. 3.28 The front view of point K.

In Fig. 3.28(a), the given two views show that the object is a partial torus of which the axis is perpendicular to plane W. Thus, its latitude-circles must be parallel to plane W. Besides, from k'', it is known that point K is on the front-upper torus surface.

Graphic solution is as follows.

Step 1 In the left view, with the intersection point of the two centerlines as center and the distance between the intersection point and the projection point k'' as radius, construct a circle which is the true shape of the latitude-circle. The latitude-circle intersects the rear-front symmetrical center line at $1''$ and $2''$. Note that the symmetrical centerline represents the front-rear symmetrical plane in space.

Step 2   From 1″ and 2″, draw their projection lines to the front view to intersect the frontal outlines at 1′ and 2′. Segment line 1′ 2′ is the front view of the latitude-circle.

Step 3   From $k″$, draw its projection line to the front view to intersect line segment 1′ 2′ at $k′$, the front view of point K (Fig. 3.28(b)).

3.2.5 Composite revolution（组合回转体）

A composite revolution is a solid with a composite revolutionary surface（组合的回转面） generated by a composite generatrix（组合的母线） sweeping around an axis. Thus, a composite revolution is also a co-axis revolution. In Fig. 3.29(a), the composite revolution is composed of a frustum cone, a small cylinder and a big cylinder. The composite revolution in Fig. 3.26(b) is composed of a small frustum cone, a small cylinder, a partial torus, a big cone, a big cylinder and a partial sphere.

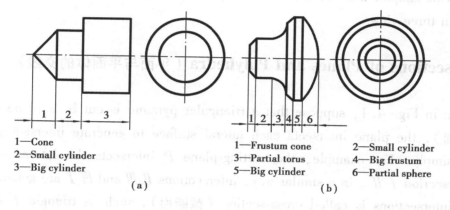

1—Cone
2—Small cylinder
3—Big cylinder

(a)

1—Frustum cone   2—Small cylinder
3—Partial torus   4—Big frustum
5—Big cylinder   6—Partial sphere

(b)

Fig. 3.29 Composite revolutions.

When drawing a composite revolution, it is required to draw each part of it respectively. If two primary revolutions are tangent, there is no border line between them because tangent joining means smooth joining. For example, in Fig. 3.29(b), the small cylinder is tangent to the partial torus, there is no border line between them. There is always a border line between two adjacent primary revolutions except when they are tangent. For example in Fig. 3.29(b), the boundary of the big frustum cone and the big cylinder should be drawn.

Chapter 4 Surface Intersections

Intersections occur very frequently in building construction, machine parts, and the designers must know how to construct them.

As we know, a line intersects a surface to generate an intersection point. However, if two surfaces intersect, the resulting line of intersection may be a straight line or a curve, or a combination of a straight line and a curve. In engineering, the accurate representation of the intersecting surfaces is very important since precise assembly is necessary for function and appearance. This chapter will cover the basic types of intersections and representing ways of the intersections in three views.

4.1 Intersections of Planes and Polyhedra（平面与平面体的交线）

As shown in Fig. 4.1, suppose that a triangular pyramid is cut by a plane called cutting-plane（截平面）, the plane intersects each lateral surface to generate intersection line, called intersection simply. For example, the cutting-plane P intersects the lateral surface SAB to generate intersection $I\,II$. In a similar way, intersections $II\,III$ and $III\,I$ are generated. The area enclosed by intersections is called cross-section（截断面）, such as triangle $I\,II$ as shown in Fig. 4.1.

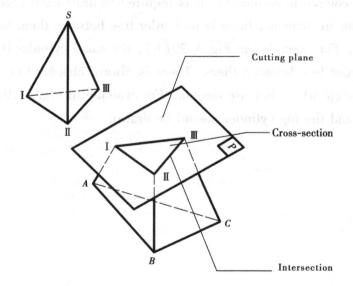

Fig. 4.1 Intersection of a cutting-plane and a polyhedron.

In plane geometry, it is known that any two planes either must be parallel or they must intersectant, even if the intersection falls beyond the limites of the plane as given. The intersection of two planes is a straight line common to the planes and its position is therefore determined by any two points common to both planes. In fact, to make the views of intersections is to find the views of common points.

Example 4.1 In Fig. 4.2(a), if a rectangular pyramid's top is cut off by a cutting-plane, complete the top and left views of the reserved (保留) portion.

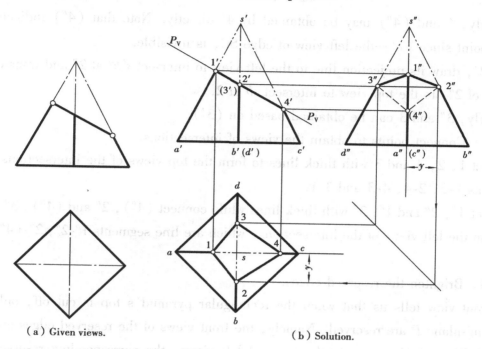

(a) Given views.　　　　(b) Solution.

Fig. 4.2　Intersections of a plane and a rectangular pyramid.

Analysis is as follows.

First of all, draw the left view of the uncut rectangular pyramid with phantom lines (假想线) and analyze views of the intersections.

Capital letter S is assigned to the vertex and capital letters, A, B, C and D are assigned to the base of the rectangular pyramid. Three views of these points are shown in Fig. 4.2(b). Spatial analysis of all surfaces is as follows. Base ABCD is parallel to plane H; other lateral surfaces including surfaces SAB, SBC, SCD and SDA are oblique planes.

The given views show that the rectangular pyramid is cut by a cutting-plane perpendicular to plane V. When capital letter P is assigned to the cutting-plane, Pv is its frontal piercing line. In other words, cutting-plane P appears as a line, Pv, in the front view. The cutting-plane P intersects respectively four lateral surfaces of the rectangular pyramid to generate four intersections which appear as P_v in the front view since intersections are common lines.

In addition because all edges of the rectangular pyramid are cut off, four edges including edges SA, SB, SC and SD intersect cutting-plane P to generate four common points whose front views are 1′, 2′, (3′) and 4′. Note that (3′) is invisible since the front view of s′d′-edge SD, is invisible. The front views of the intersections are straight line segments composed by these projection points, which are line segments 1′-2′, 2′-4′, 4′-(3′) and (3′)-4′.

To construct other views of these intersections is actually to complete other views of these common points in terms of their front views, 1′, 2′, (3′) and 4′.

Graphic solution is as follows.

Step 1 Complete other views of the common points.

From 1', draw its projection line to the top and left views to intersect sa and $s''a''$ at 1 and 1'', respectively.

Similarly, 4 and (4'') may be obtained by 4' directly. Note that (4'') indicates invisible projection point since $s'c'$—the left view of edge SC, is invisible.

From 2', draw its projection line to the left view to intersect $s''b''$ at 2'' and transfer y—the Y coordinates of 2'', to the top view to intersect sb at 2.

Similarly, 3'' and 3 can be obtained based on (3').

Step 2 Connect points to obtain the views of intersections.

Connect 1, 2, 4 and 3 with thick lines to form the top views of the intersections, which are line segments 1-2, 2-4, 4-3 and 3-1.

Connect 1'', 2'' and 1'', 3'' with thick lines while connect (4''), 2'' and (4''), 3'' with hidden lines to form the left views of the intersections, which are line segments 1''-2'', 2''-(4''), (4'')-3'' and 3''-1''.

Step 3 Brighten the reserved edges.

The front view tells us that when the rectangular pyramid's top is cut off, only the edges below cutting-plane P are reserved. Namely, the front views of the reserved edges are 1'-a', 2'-b', (3')-(d') and 4'-c', so in the top and left views, the corresponding reserved edges are brightened, which are the line segments 1-a, 2-b, 4-c, 3-d and 1''-a'', 2''-b'', 3''-d''. Besides, hidden line (4'')-(c'') coincides with thick line 1''-a'', so it is drawn by a thick line.

Lastly, it is necessary to check, clean and brighten all the thick lines.

Example 4.2 Fig. 4.3 (a) shows that the triangular pyramid is cut by two cutting-planes. Complete the top and left views of the reserved portion of the object.

Analysis is as follows.

First of all, draw the left view of the uncut triangular pyramid with phantom lines and analyze views of the intersection.

Capital letter S is assigned to the vertex and capital letters, A, B and C are assigned to the base of the triangular pyramid. Three views of these points are shown in Fig. 4.3(b). Spatial positions of all the surfaces are as follows. Base ABC is parallel to plane H. Lateral surfaces SAB, SAC are oblique planes while lateral surface SBC is perpendicular to plane V since line BC appears as a point $b'(c')$ in plane V.

The given views show that triangular pyramid is cut by two cutting-planes in which one is perpendicular to plane V and the other is parallel to plane H. When capital letter Q is assigned to the cutting-plane perpendicular to plane V, Qv is its frontal piercing line. In other words, cutting-plane Q appears as a line, Qv, in the front view. The intersections of cutting-plane Q and each lateral surface of the triangular pyramid are converged on Qv in the front view since intersections are common lines.

Chapter 4 Surface Intersections

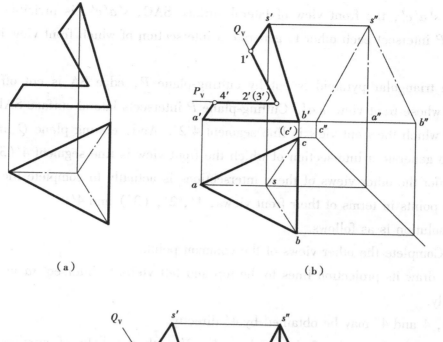

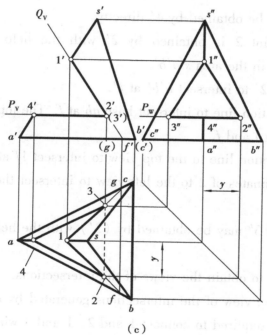

Fig. 4.3 A triangular pyramid is cut by two cutting planes.

When capital letter P is assigned to the horizontal cutting-plane, P_V is its frontal piercing line and P_W is its profile piercing line. In other words, the cutting-plane P appears as a line, P_V, in the front view and as a line, P_W, in the left view, respectively. The intersections of cutting-plane P and each lateral surface of the triangular pyramid are converged on P_V since the intersections are also common lines.

When the triangular pyramid is cut by cutting-plane Q, edge SA is cut off to generate a common point whose front view is 1′. cutting-plane Q intersects lateral surface SAB to generate a intersection whose front view is line segment 1′2′. And cutting-plane Q intersects lateral surface SAC to generate a intersection whose front view is line segment 1′(3′). Note that (3′) is

77

invisible since $s'a'c'$, the front view of lateral surface SAC, $s'a'c'$, is invisible. Two cutting-planes Q and P intersect each other to generate a intersection of which front view is line segment $2'(3')$.

When the triangular pyramid is cut by cutting-plane P, edge SA is cut off to generate a common point whose front view is $4'$. Cutting-plane P intersects lateral surface SAB to generate a intersection of which the front view is line segment $4'2'$. And, cutting plane Q intersects lateral surface SAC to generate a intersection of which the front view is line segment $4'(3')$.

To construct the other views of these intersections is actually to complete the other views of these common points in terms of their front views, $1'$, $2'$, $(3')$ and $4'$.

Graphic solution is as follows.

Step 1 Complete the other views of the common points.

From $1'$, draw its projection lines to the top and left views to intersect sa and $s''a''$ at 1 and $1''$, respectively.

Similarly, 4 and $4''$ may be obtained by $4'$ directly.

However, projection point 2 is obtained by $2'$ with the help of auxiliary line because projection point $2'$ is located in the area $s'a'b'$.

Extend line segment $1'-2'$ to intersect $a'b'$ at f'.

From f', draw its projection line to intersect base ab at f. The top view of the auxiliary line, 1f, is obtained by connecting 1 and f.

From $2'$, draw its projection line to the top view to intersect 1f at 2.

Transfer y—the Y coordinates of 2 to the left view to intersect the other projection line from $2'$ at $2''$.

In the same way, 3 and $3''$ may be obtained by $(3')$ with the help of the auxiliary line $(1'-(g'), 1-g)$.

Step 2 Connect points to obtain the views of the intersections.

In order to obtain the top view of the intersections generated by cutting-plane Q intersecting the triangular pyramid, it is required to connect 1 and 2, 1 and 3 with thick lines while 2 and 3 with a hidden line because projected from upper to bottom, the line segment 23 is invisible. It is also required to connect $1''$ and $2''$, $2''$ and $3''$, $3''$ and $1''$ with thick lines to form the left views of the intersections, also generated by cutting-plane Q intersecting the triangular pyramid.

Similarly connect 4 and 2, 4 and 3 with thick lines to form the top views of the intersections. generated by cutting-plane P intersecting the triangular pyramid. The left views of the intersections appear as edges on P_W, line segments $2''-4''$ and $4''-3''$.

Step 3 Brighten the reserved edges.

The front view tells us that when the tetrahedron is cut by cutting-planes Q and P, the edge between the two cutting-planes is cut off. The front views of the reserved edges are $1'-s'$, $4'-a'$. Therefore, in the top and left views, corresponding edges, 1-s, 4-a, and $1''-s''$, $4''-a''$ are brightened. Besides, the front view also shows that edges SB and SC are not cut off. That is to say, $s'b'$ and $s'(c')$ are complete. The edges, sb and sc in the top view as well as $s''b''$ and $s''c''$ in the left view are brightened.

Lastly, check, clean and brighten all the thick lines to complete the top and left views of the reserved tetrahedron.

Example 4.3 Fig. 4.4(b) shows that a rectangular prism is cut by two cutting-planes. Complete the top view and supplement the left view of the reserved portion.

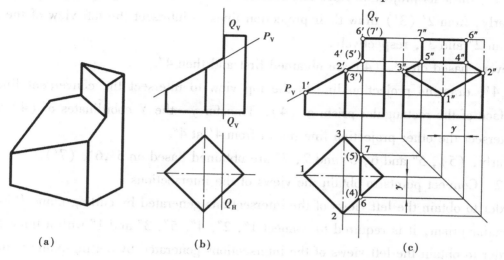

Fig. 4.4 Intersections of two cutting-planes and a rectangular prism.

Analysis is as follows.

First of all, draw the left view of the uncut rectangular prism with phantom lines and analyze the views of the intersections.

The given views show that four lateral surfaces of the rectangular prism are perpendicular to plane H while its two bases are parallel to plane H.

Capital letter P is assigned to the cutting plane perpendicular to plane V, thus P_V is its frontal piercing line. In other words, cutting-plane P appears as a line, P_V, in the front view. The front views of the intersections of the cutting-plane P and each lateral surface of the rectangular prism are converged on P_V because the intersections are common lines. When capital letter Q is assigned to the profile cutting-plane, Q_V is its frontal piercing line and Q_H is its horizontal piercing line. In other words, the cutting-plane Q appears as a line, Q_v, in the front view and as a line, Q_H, in top view, respectively. The front views of the intersections of the cutting-plane Q and each lateral surface of the rectangular prism are converged on Q_V since the intersections are also common lines.

When the rectangular prism is cut by cutting-plane P, its left edge is cut off to generate a common point whose front view is $1'$. Its front and rear edges are also cut off to generate two common points whose front views are $2'$ and $(3')$. The two cutting-planes intersect to generate a intersection whose front view is line segments $4'$-$(5')$.

When the rectangular prism is cut by cutting-plane Q, the cutting-plane Q intersects the right-front and right-rear lateral surfaces of the rectangular prism to generate intersections whose front views are line segments 6'-4' and (7')-(5').

To construct other views of these intersections is actually to complete other views of these common points in terms of their front views, 1', 2', 4', (3'), (5'), 6', and (7').

Graphic solution is as follows.

Step 1 Complete the other views of the common points.

From 1', draw its projection lines to intersect the left view of the left edge at 1″.

Similarly, from 2' (3') draw their projection lines to intersect the left view of the front and rear edges at 2″ and 3″, respectively.

However, based on 4', (4) are obtained first and then 4″.

From 4', draw its projection line to the top view to intersect the convergent line of front lateral surface of the rectangular prism at (4). Transfer y—the Y coordinates of (4) to the left view to intersect the other projection line drawn from 4' at 4″.

Similarly, (5), 5″ and 6, 6″ and 7, 7″ are obtained based on 5',6', (7').

Step 2 Connect points to obtain the views of the intersections.

In order to obtain the left views of the intersections generated by cutting-plane P intersecting the rectangular prism, it is required to connect 1″, 2″, 4″, 5″, 3″ and 1″ with a thick lines.

In order to obtain the left views of the intersections generated by cutting-plane Q intersecting the rectangular prism, it is also required to connect 6″, 4″, 5″, 7″ and 6″ with thick lines.

Completes the top view of the object by connecting 6 and 7 with a thick line.

Step 3 Brighten the reserved edges.

The front view shows that the portion above cutting-plane P and to the left of cutting-plane Q is cut off. Thus, the reserved edges are below P_v and to the right of Q_v. In the left view, the edges below 1″, 2″ and 3″ are brightened. Besides, the edge above 1″ is drawn with a hidden line as the left view of the right edge is invisible.

Lastly, check, clean and brighten all the thick lines to complete the top and left views of the reserved part of the rectangular prism.

4.2 Intersections of Planes and Revolutions (平面与回转体的交线)

Incomplete revolutions occur frequently in engineering practice as shown in Figs. 4.5(a), (b) and (c), so intersections exist on the revolutionary surfaces. Shape of intersections depends on the shape of the revolution and the location of the cutting-plane relative to the axis of the revolution. Intersections are straight line segments or curves. If it is a curve, a sufficient number of points must be located to obtain an accurate intersection.

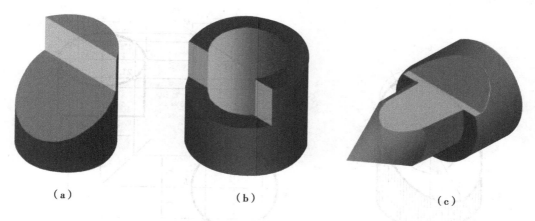

Fig. 4.5 Incomplete revolutions.

4.2.1 Intersection of a plane and a cylinder

Chart 4.1 shows three possible cases. When the cutting-plane is perpendicular to the axis of the cylinder, the intersection is a circle as shown in Chart 4.1 Column 1. When the cutting-plane is oblique to the axis of the cylinder, the intersection is an ellipse as shown in Chart 4.1 Column 2. When the cutting-plane is parallel to the axis of the cylinder, the intersections are two elements as shown in Chart 4.1 Column 3.

Chart 4.1 Intersections of a plane and a cylinder

Locations of cutting plane	Perpendicular to axis	Oblique to axis	Parallel to axis
Intersection	Circle	Ellipse(椭圆)	Two elements
Pictorial drawing(直观图)			
Three views			

Example 4.4 Fig. 4.6 shows an incomplete cylinder. Complete the left view of the cylinder.

Analysis is as follows.

First of all, draw the left view of the cylinder with phantom lines and analyze the views of the

Fundamentals of Engineering Drawing

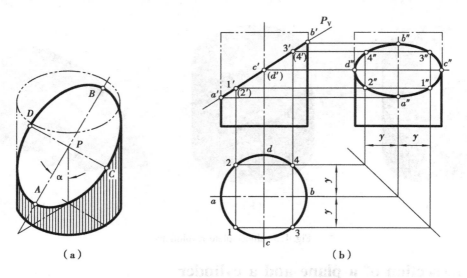

Fig. 4.6 Intersection of a plane and a cylinder.

intersections.

The given views show that the incomplete cylinder is formed by a cutting-plane cutting a cylinder. If a capital letter P is assigned to the cutting-plane, P_v is its frontal piercing line. In other words, cutting-plane P appears as a line, P_v, in the front view. The front view of the intersection is located on P_v since the intersection is a common line. On the other hand, because the cutting-plane P is oblique to the axis of the cylinder, the intersection is an ellipse in the space. The ellipse appears as an edge on P_V in the front view. The ellipse's top view is a circumference because each point on the ellipse is on the cylindrical surface. Only the left view of the ellipse needs to be shown.

It is known that in plane geometry, there are two axes on the ellipse, which are major axis AB and minor axis CD as shown in Fig. 4.6(a). The major and minor axes intersect perpendicularly and bisect (平分) each other. Endpoints on the major and minor axes are special common points on the intersection. P_v intersects two frontal outlines of the cylinder at a' and b', which are the front views of the major axis' endpoints; while P_v intersects the axis of the cylinder at $c'(d')$, which are the front views of minor axis' endpoints. On P_v, two pairs of projection points, $1'(2')$ and $3'(4')$, are picked up properly, which are front views of the two pairs of general common points on the ellipse. So, to complete the left view of the intersection is actually to complete the left view of these common points.

Graphic solution is as follows.

Step 1 Complete other views of the common points.

It is known from 3.2.2 that revolution's outlines are always located on the symmetrical plane. For example, profile outlines are located on left-right symmetrical plane while frontal outlines are located on front-rear symmetrical plane.

From a' and b', draw their projection lines to the left view to intersect the front-rear symmetric plane at a'' and b''.

From $c'(d')$, draw its projection line to the left view to intersect the profile outline at c'' and d''.

From $1'$, $(2')$, draw their projection line to the top view to intersect the circumference at 1 and 2. Transfer y—the Y coordinates of 1 and 2, to the left view to intersect the other projection line drawn from $1'(2')$ at $1''$ and $2''$, respectively.

Similarly, $3''$ and $4''$ may be obtained based on $3'$, $(4')$.

Step 2 Connect points smoothly to obtain the views of the intersections.

In order to obtain the left view of the intersections generated by cutting-plane P intersecting the cylinder, connect a'', $1''$, c'', $3''$, b'', $4''$, d'', $2''$ and a'' smoothly with thick lines.

Step 3 Brighten the reserved profile outlines.

The front view shows that the portion above cutting-plane P is cut off, so the reserved profile outlines are below the P_V. In other words, the reserved profile outlines are below $c'(d')$. In the left view, the profile outlines below c'' and d'' are brightened. Note that the center line below a'' is not brightened because the cylindrical surface is smooth.

Lastly, check, clean and brighten all the thick lines to obtain the left view of the object.

Example 4.5 Fig. 4.7 shows a cylinder with a slot (槽). Supplement the left view of the object.

Analysis 1 is as follows.

First of all, draw the left view of the complete cylinder with phantom line and analyze the views of the intersections.

The given views show that a cylinder, whose axis is perpendicular to plane H, is cut by three cutting-planes to form a slot. Two side surfaces of the slot are parallel to plane W and the bottom base is parallel to plane H.

When capital letters P_1, P_2 are assigned to the left side surface and right side surface of the slot, P_{1v} and P_{2v} are their frontal piercing lines while P_{1H} and P_{2H} are their horizontal piercing lines. Evidently, cutting-planes P_1 and P_2 are not only parallel to axis of the cylinder but also parallel to plane W, so the intersections are four elements on the cylinder. The top views of the intersections are located on the P_{1H} and P_{2H}, and their front views are on P_{1v} and P_{2v} since intersections are common lines. Only the left views of the intersections need to be shown. In addition, cutting-planes P_1 and cutting-plane P_2 are symmetric to the axis of the cylinder, so, only the intersections generated by one of two cutting-planes, such as cutting-plane P_1, intersecting the cylinder sleeve, need to be discussed. P_{1H} intersects circumference at $a(b)$ and $g(h)$, which are the top views of the intersections AB and GH as shown in Fig. 4.7(a). The front views of the two intersections are on P_{1v}, that is, $a'b'$ and $(g')(h')$.

To show the left views of these intersections is actually to add the left view of these common points by means of their two known views.

Graphic solution 1 is as follows.

To obtain a'', it is required to transfer y—the Y coordinates of a, to the left view through 45° miter-line to intersect the other projection line drawn from a' at a''. In the same way, b'', g'' and h'' can be obtained. Connecting a'' and b'', g'' and h'' with thick lines completes the left views of

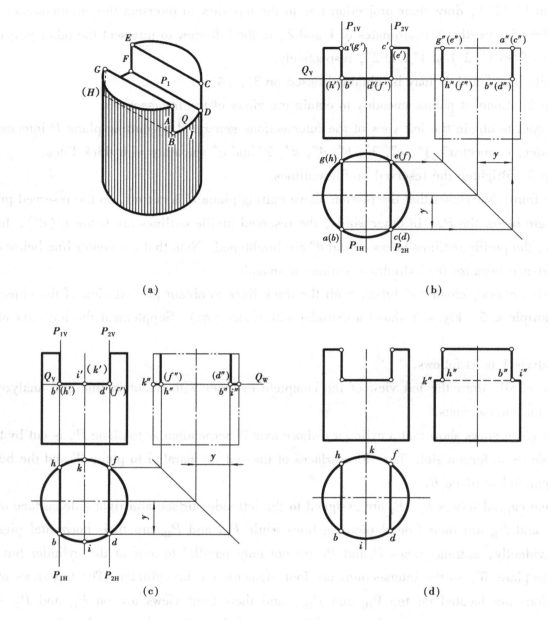

Fig. 4.7 Intersections of multi-planes and a cylinder.

the intersections (Fig. 4.7(b)).

Analysis 2 is as follows.

When capital letter Q is assigned to the bottom base of the slot, Q_v and Q_w are their frontal and left piercing lines, respectively. Cutting-plane Q is perpendicular to the axis of the cylinder, so the intersections are two circular arcs in the space and their top view are two arcs, b-i-d and h-k-f, on the circumference as shown in Fig. 4.7(c). The front views of the intersections appear as line segments on the Q_v, b'-i'-d' and (h')-(k')-(f'). Only the left views of the intersections need to be shown.

Graphic solution 2 is as follows.

Because top projection points b and d have the same Y coordinates, when they are projected from left to right, (d'') coincides with b''. For the same reason, (f'') coincides with h''. Since top

84

projection points i and k are located on the left-right symmetric plane, their left views, i'' and k'', are located on profile outlines of the cylinder. Connect b'' and i'', h'' and k'' with thick lines to complete the left views of the intersection. Moreover, horizontal cross-section, b-i-d-f-k-h, represents true size of the bottom base of the slot and its left view appears as line segments on Q_W. On Q_W, the line segment $b''h''$ is invisible since when the cylinder is projected from left to right, the middle partial bottom of the slot is hidden by left partial cylinder Fig. 4.7(d).

And then, brighten the reserved profile outlines. The front view shows that the portion above cutting-plane Q is cut off, so the reserved profile outlines are below the Q_V. In other words, the reserved profile outlines are below $i'(k')$ as shown in Fig. 4.7(c). The profile outlines below i'' and k'' are brightened as shown in Fig. 4.7(d).

Lastly, check, clean and brighten all the thick lines to obtain the left view of the cylinder with a slot.

Example 4.6 The cylinder sleeve is cut by three cutting-planes to form two slots as shown in Fig. 4.8(a). Complete the left view of the object.

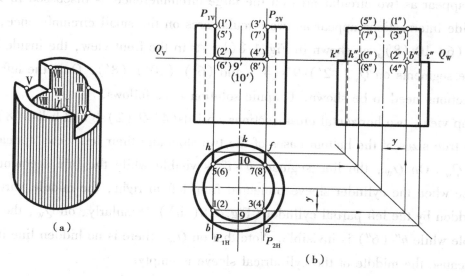

Fig. 4.8 Intersections of a cylinder sleeve and three cutting planes.

Fig. 4.8 is similar to Fig. 4.7. The difference between them is that Fig. 4.8 shows a cylinder sleeve (圆筒), so there are two cylindrical surfaces, i.e. exterior and interior cylindrical surfaces. The top view of the exterior cylindrical surface is a large circumference while the interior cylindrical surface is a small circumference. The intersection generated by cutting-plane intersecting the exterior cylindrical surface is the outer intersection while the intersection generated by cutting-plane intersecting the interior cylindrical surface is inner intersection.

Analysis on three cutting-planes is the same with the previous discussion. When the cylinder sleeve is cut by cutting-planes P_1 and P_2, besides four outer intersections, there are also four inner intersections. As the four outer intersections have been discussed in Fig. 4.7, now only four inner intersections need to be discussed. P_{1V}, P_{2V}, P_{1H}, P_{2H} in Fig. 4.8 are the same with those in Fig. 4.7. The top and front views of the intersections are located on P_{1H} and P_{2H}, P_{1V} and P_{2V}, respectively. Only the left views of the intersections need to be shown. Furthermore cutting-plane

P_1 and cutting-plane P_2 are symmetric to the axis of the cylinder sleeve, so, only the intersections generated by one of the two cutting-planes such as cutting-plane P_1 intersecting the cylinder sleeve, need to be discussed. P_{1H} intersects the small circumference at pointes 1(2) and 5(6), which are the top views of the two inner intersections. The front views of the two inner intersections are line segments (1')-(2') and (5')-(6') on P_{1V}. The left views of these intersections are shown as follows.

y—The Y coordinates of the top projection point 1 is transferred to the left view to intersect the other projection line drawn from (1') at (1"). Similarly, (2"), (5"), (6") may be obtained. Connect (1") and (2"), (5") and (6") with hidden lines to form the left views of inner intersections.

When the cylindrical sleeve is cut by cutting-plane Q, intersections are four circular arcs, two of which are located on the outside cylindrical surface—outside intersections while the other two are located on the Inside cylindrical surface—inside intersections. in the top view, the outside intersections appear as two circular arcs on the large circumference as discussed in Fig. 4.7(b) while the inside intersections appear as two circular arcs on the small circumference, arc (2)-9-(4) and arc (6)-10-(8), as shown in Fig. 4.8(b)). In the front view, the inside intersections appear as line segments on Q_v, (2')-9'-(4') and (6')-(10')-(8'). Only the left views of the inside intersections need to be shown. Graphic solution is as follows.

In the top view, two horizontal cross-sections, b-i-d-(4)-9-(2)-b and h-k-f-(8)-10-(6)-h, represent the true sizes of the bottom bases of the two slots and their left views appear as two line segments on Q_w. On Q_W, the line segment b''-i'' is visible while the line segment (2")-b'' is invisible since when the cylinder sleeve projected from left to right, the middle partial bottom of the slot is hidden by the left partial cylinder (Fig. 4.8(b)). Similarly, on Q_W, the line segment k''-h'' is visible while h''-(6") is invisible. Note that on Q_W, there is no hidden line between (2") and (6") because the middle of the cylindrical sleeve is empty.

Finally, check, clean and brighten all the thick lines to form the left view of the cylinder with two slots.

4.2.2 Intersections of a plane and a cone

Chart 4.2 shows five types of intersections of a cone and a cutting-plane, according to the different positions of the cutting-plane.

When a cone is cut by planes at different angles, five intersections are obtained.

When the cutting-plane is perpendicular to the axis of the cone, the intersection is a latitude-circle as shown in Chart 4.2 Row 1. When the cutting-plane makes a greater angle with the axis than a half vertex angle of the cone, i.e. $\beta > \varphi$, the intersection is an ellipse as shown in Chart 4.2 Row 2. When the cutting-plane makes the same angle with axis as the half vertex angle of the cone, i.e. $\beta = \varphi$, the intersection is a parabola as shown in Chart 4.2 Row 3. When the cutting-plane makes smaller angle with the axis than the half vertex angle and it is parallel to the axis, i.

e. $\beta < \varphi$, the intersection is a hyperbola as shown in Chart 4.2 Row 4. Finally, if the cutting-plane passes through the vertex, intersections are two elements as shown in Chart 4.2 Row 5.

Two methods are employed to find the views of the intersections, which are element-line method (素线法) and latitude-circle method (纬圆法).

Chart 4.2 Intersections of a cutting-plane and a cone

Location of the cutting plane	Pictorial drawing	Three views
□$P \perp OO$	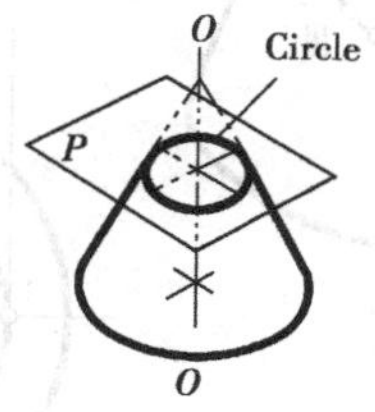 Circle	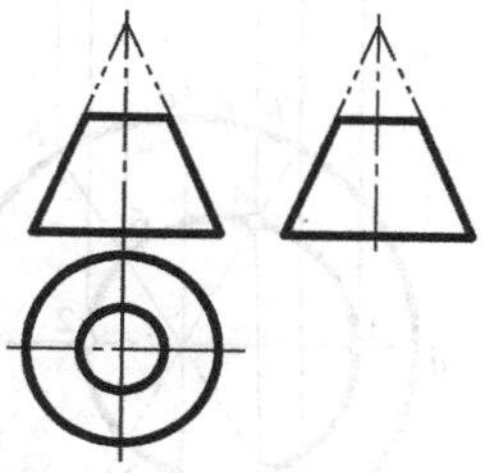
□$P \cap OO$ and $\beta > \varphi$	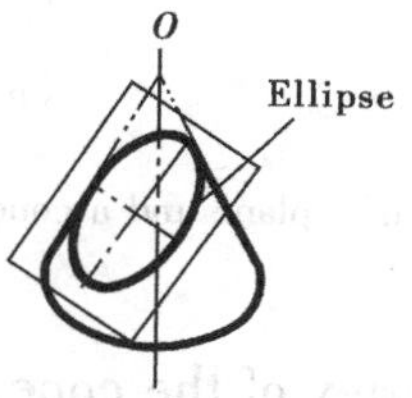 Ellipse	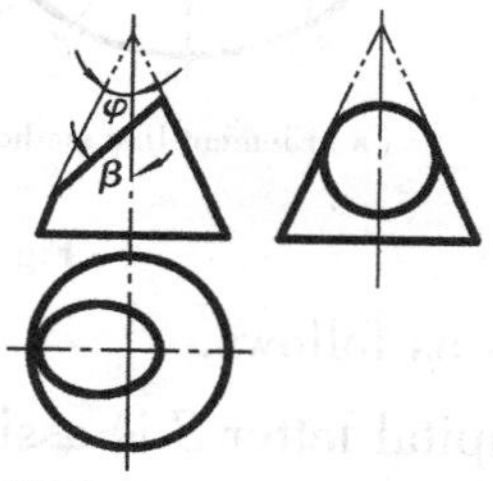
P//element line of the cone and $\beta = \varphi$	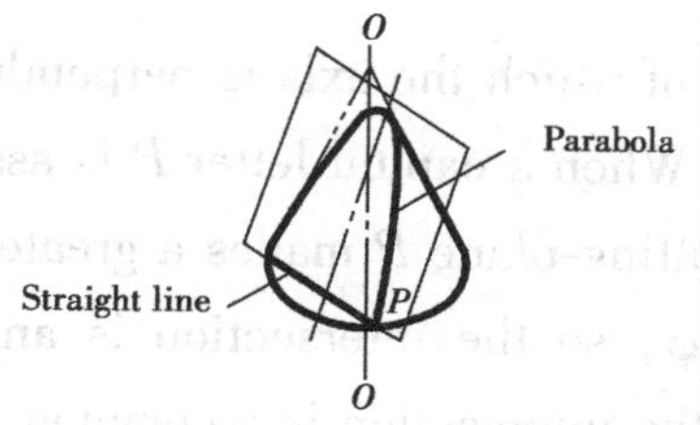 Parabola, Straight line	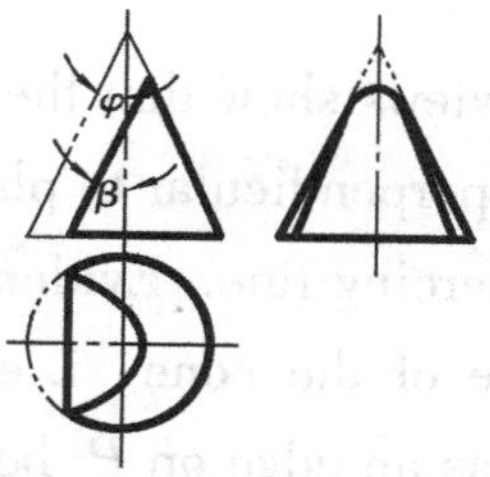
□P//OO and $\beta < \varphi$	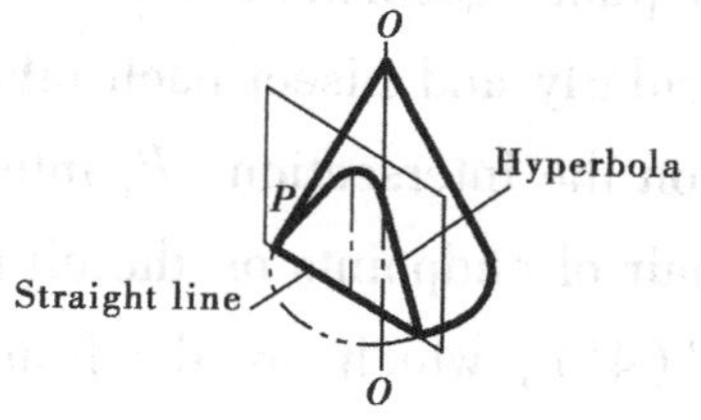 Hyperbola, Straight line	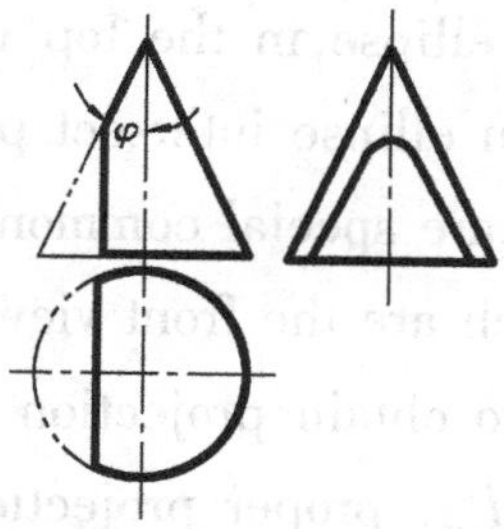
□P passes through the vertex of the cone	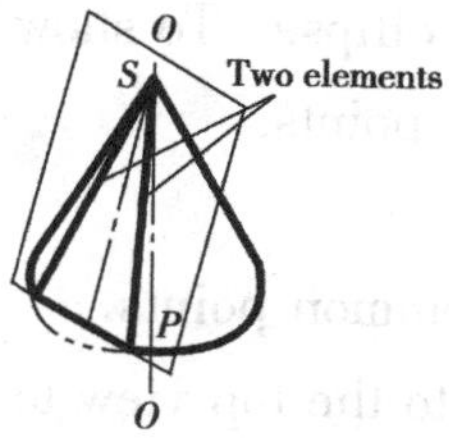 Two elements	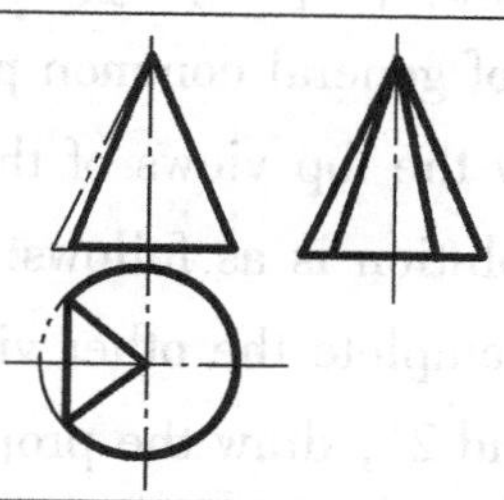

Example 4.7 Fig. 4.9 shows that a cone' top is cut off. Complete the top view of the object.

Fig. 4.9(a) illustrates the process to find the view of the intersection by means of element-line method.

87

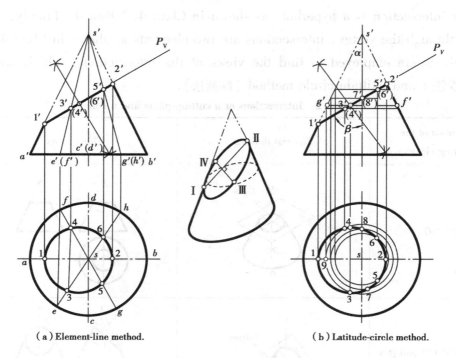

(a) Element-line method.　　(b) Latitude-circle method.

Fig. 4.9　Intersection of a plane and a cone.

Analysis is as follows.

When a capital letter S is assigned to the vertex of the cone, s and s' are the two views of the vertex.

The given views show that the cone, of which the axis is perpendicular to plane H, is cut by a cutting-plane perpendicular to plane V. When a capital letter P is assigned to the cutting-plane, P_v is its front piercing line. Evidently, cutting-plane P makes a greater angle with the axis than a half vertex angle of the cone, i. e. $\beta > \varphi$, so the intersection is an ellipse in the space. The ellipse appears as an edge on P_v because the intersection is a common line. The ellipse appears as a foreshortened ellipse in the top view. In plane geometry, it is known that the major and the minor axes of an ellipse intersect perpendicularly and bisect each other. Endpoints on the major and minor axes are special common points on the intersection. P_v intersects the frontal outlines at $1'$ and $2'$, which are the front views of a pair of endpoints on the ellipse. Bisect (平分) the line segment $1'$-$2'$ to obtain projection point $3'(4')$, which are the front views of the other pair of endpoints. On P_v, proper projection points such as $5'(6')$ are picked up, which are the front views of a pair of general common points on the ellipse. To show the top view of the intersection is actually to show the top views of these common points.

Graphic solution is as follows.

Step 1　Complete the other view of the common points.

From $1'$ and $2'$, draw the projection lines to the top view to intersect the front-rear symmetric plane at 1 and 2, respectively.

Take $3'$ as an example to illustrate how to find its top views by means of element-line method.

Connect s' and $3'$ to obtain line segment s'-$3'$ and extend it to intersect the bottom of the cone at e'. Connect s' and e' to obtain the line segment s'-e'.

Through e', draw its projection line to the top view to intersect the bottom circle at e, and connect s and e to obtain the line segment se.

The two views, se and $s'e'$, represent the element-line on the conical surface through projection point $3'$.

Through $3'$, draw its projection line to the top view to intersect se at 3.

In the same way, a series of top projection points such as 4, 5, 6, are gotten.

Step 2 Connect points smoothly to form the top view of the intersection.

In the top view, each projection point is connected with a thick line to show the top view of the intersection, ellipse 1-3-5-2-6-4-1.

Lastly, check, clean and brighten all the thick lines. So far, the top view of the portion of the cone is completed.

Fig. 4.9(b) illustrates the process to find the view of the intersection by means of the latitude-circle method.

It is known from 3.2.1 that any latitude-circle is always perpendicular to the axis of a revolution. In Fig. 4.9(b), latitude-circles are a series of horizontal circles since the axis of the given cone is perpendicular to plane H, and the front view of any latitude-circle in the problem appears as an edge perpendicular to the axis and its top view is actually a circle. Take $3'(4')$ as an example to illustrate how to find their top views by means of latitude-circle method.

Through $3'(4')$, draw a line perpendicular to the axis of the cone, which intersects outlines at g' and f'.

Through g', draw its projection line to the top view to intersect the front-rear symmetric plane at g.

With line segment sg as radius and cone's vertex s as the center, draw latitude-circle to intersect the other projection line drawn from $3'(4')$ at 3 and 4 respectively.

In the same way, a series of top projection points such as 5, 6, are gotten.

Furthermore, the method to find top projection points, 1 and 2, is the same as stated in Fig. 4.9(a).

Lastly, connect each top projection points smoothly with a thick line to show the top view of the intersection, ellipse 1-3-7-5-2-6-8-4-1.

4.2.3 Intersections of a plane and a sphere

If a sphere is cut by a cutting-plane, the intersection is always a circle in space. When the cutting-plane is parallel to a certain projection plane, the intersection appears as a circle in this projection plane and as edges in other projection planes. The length of the edge equals the diameter of the circle. For example, in Fig. 4.10, since cutting-plane Q is parallel to plane H, the intersection is a horizontal circle in space. The circle appears as a circle, q, in the top view while as edges, q' and q'', in the front and left views, respectively. Similarly, since cutting-plane

P is parallel to plane W, the intersection is a profile circle in space. The circle appears as a circle, p'', in the left view while as edges, p' and p, in the front and top views, respectively.

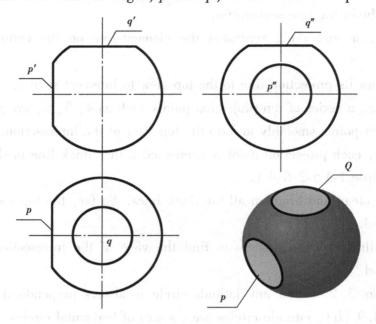

Fig. 4.10 Intersections on a sphere.

When the cutting-plane is perpendicular to a certain projection plane, the intersection appears as an oblique edge in this projection plane and as ellipses in other views.

Example 4.8 As shown in Fig. 4.11, a sphere is cut by a cutting-plane. Complete the top view of the reserved portion.

Analysis is as follows.

First of all, draw the top view of the uncut sphere with a phantom line and analyze the views of the intersection.

Two views of the sphere, center are o' and o.

The given views show that the cutting-plane is perpendicular to plane V. When the capital letter P is assigned to the cutting-plane, P_v is its frontal piercing line. The intersection of cutting-plane P and the spherical surface is a circle in space. However, the circle appears as an edge on P_v in the front view and as an ellipse in the top view. Endpoints on the major and minor axes of the ellipse are special common points on the intersection. P_v intersects the frontal outline at a' and b', which are the front views of a pair of endpoints on the ellipse. Bisect $a'b'$ to obtain $c'(d')$, which are the front views of the other pair of endpoints. P_v intersects the horizontal outline at $1'$ $(2')$, which are front views of a pair of special common points on the horizontal outline of the sphere. On P_v, pick up proper projection points such as $3'(4')$ which are the front views of a pair of general common points on the intersection. Therefore, to show the top view of the intersection is actually to show the top views of these common points.

Graphic solution is as follows.

Step 1 Complete the top view of the the common points.

From a' and b', draw projection lines to intersect the front-rear symmetric plane at a and b,

Chapter 4 Surface Intersections

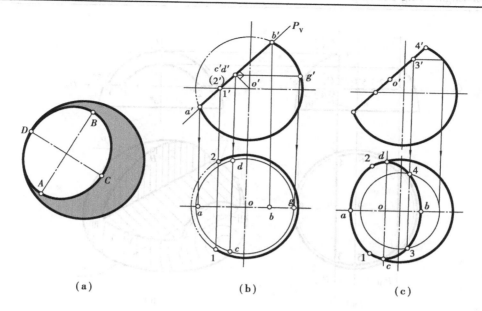

(a) (b) (c)

Fig. 4.11 Intersection of a sphere and the cutting-plane P.

respectively (Fig. 4.11(b)).

From $1'(2')$, draw their projection lines to intersect the horizontal outline of the sphere at 1 and 2 (Fig. 4.11(b)).

Through $c'(d')$, draw a line perpendicular to the axis of the sphere to intersect the frontal outline at g'. From g', draw its projection line to intersect the front-rear symmetric plane at g (Fig. 4.11(c)).

With line segment og as radius and the center of the sphere, o, as center draw a latitude-circle to intersect the other projection drawn line from $c'(d')$ at c and d, respectively.

In the same way, a series of top projection points such as 3 and 4 can be obtained.

Step 2 Connect points smoothly to form the view of the intersection.

Connect each top projection points smoothly with a thick line to show the top view of the intersection, ellipse a-1-c-3-b-4-d-2-a.

Step 3 Brighten the reserved outlines.

The front view shows that the sphere above cutting-plane P is cut off, so the reserved horizontal outline is below the P_V. In other words, the reserved horizontal outline is to the right of P_V. In the top view, brighten the horizontal outlines to the right $1''$ and $2''$ as shown in Fig. 4.11(c).

Example 4.9 Fig. 4.12 shows a hemisphere with a slot. Supplement its top view and left view.

Analysis is as follows.

First of all, draw the top and left views of the hemisphere with phantom lines and analyze the views of intersections.

The front view shows that a hemisphere is cut by three cutting-plans to form a slot.

When capital letters P_1 and P_2 are assigned to the left and right cutting-planes respectively, P_{1V} and P_{2V} are their frontal piercing lines while P_{1H} and P_{2H} are their horizontal piercing lines. Evidently, cutting-planes P_1 and P_2 are parallel to plane W. Therefore, the intersections are two

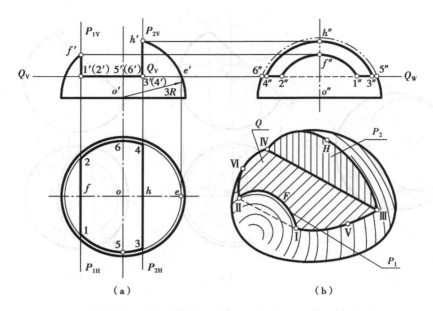

Fig. 4.12 Intersections of a sphere and multiple cutting-planes.

partial profile circles with different radius, arc I F II and arc III H IV (Fig. 4.12(b)). And when the bottom of the slot is named by a capital letter Q, Q_V is its frontal piercing line and Q_W is its profile piercing line. Obviously, cutting-plane Q is parallel to plane H, so intersections are two circular arcs on one horizontal circle, arc I V III and arc II VI IV as shown in Fig. 4.12(b).

P_{1V} and P_{2V} intersect the frontal outline of the hemisphere at f' and h'. Q_V intersects the frontal outline of the hemisphere at e' and intersects the left-right symmetric plane of the hemisphere at $5'(6')$. Q_V intersects P_{1V} and P_{2V} at $1'(2')$ and $3'(4')$.

Graphic solution is as follows.

Three views of the center of the hemisphere are o', o, o''.

Step 1 Completing the other views of intersections.

From f', draw its projection line to the left view to intersect the front-rear symmetric plane at f''.

With line segment $o''f''$ as radius and o'' as center, draw a partial profile circle to intersect Q_W at $1''$ and $2''$.

Similarly, h'' may be gotten from h'. With line segment $o''h''$ as radius and o'' as center. Draw a partial profile circle to intersect Q_W at $3''$ and $4''$.

From e', draw its projection line to the top view to intersect front-rear symmetric plane at e.

With line segment oe as radius and o as center, draw a partial horizontal circle to intersect P_{1H} at 1 and 2, P_{2H} at 3 and 4 and the left-right symmetric plane at 5 and 6. In the top view, the area 1-5-3-4-6-2-1 represents the true size of the bottom base of the slot and its left view appears as a line segment on Q_W.

On Q_W, the line segment $1''2''$ is invisible since when the hemisphere is projected from left to right, the middle partial bottom of the slot is hidden by the left partial sphere. As a result, the line segment $1''2''$ is drawn with a hidden line. In addition, line segments $2''6''$ and $1''5''$ are visible and they are drawn with thick lines (Fig. 4.12(a)).

Step 2   Brighten the reserved outlines.

The front view shows that the horizontal outline of the hemisphere is full since the upper-lower symmetric plane of a sphere is complete. The front view also shows that the portion between cutting planes P_1 and P_2 is cut off. In other words, the profile outline below Q_v is reserved. So, in the left view, the profile outline below $5''$ and $6''$ are brightened.

4.2.4 Intersections of cutting-planes and a composite revolution（平面与组合回转体的交线）

Example 4.10   Fig. 4.13 shows that a composite revolution is cut by a cutting-plane. Complete the front view of the reserved portion.

Analysis is as follows.

Two views of the sphere's center are o', o''. Point o_1 is the center of generatrix circle of the partial torus.

Let's analyze the construction of the composite revolution. Because the composite revolutionary surface is smooth, there is no edge on it. However the border line of any two adjacent primary revolutionary surfaces should be found out.

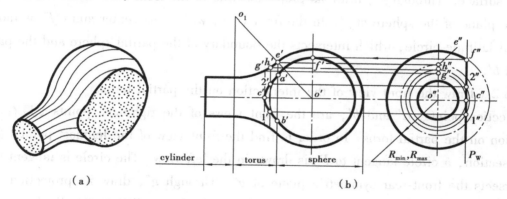

Fig. 4.13   Intersection of a combined solid of revolution with a plane.

First of all, in the front view, determine the border line between two adjacent revolutionary surfaces in the composite revolution.

Line segment $o_1 o'$ intersects the frontal outline of the composite revolution Q at e'. Through e', a line perpendicular to the axis of the composite revolution is drawn, which is the boundary of the partial torus and the partial sphere. Similarly, through o_1, a line perpendicular to the axis of the composite revolution is drawn, which is the boundary of the cylinder and the partial torus. Imaginarily, the composite revolution is disassembled in mind（分解）into three parts—the cylinder, the partial torus and the partial sphere by the two border lines (Fig. 4.13(b)).

Second, in the left view, determine the border circle between two adjacent revolutionary surfaces in the composite revolution.

Through e', draw its projection line to left view to intersect the front-rear symmetrical pane at e''. With line segment $o''e''$ as radius and o'' as center, draw a circle with a phantom line, which is the boundary circle between the partial torus and the partial sphere. In the left view, the small circumference is the projection of the cylinder. The area ranging from the small circumference to the phantom circle is the projection of the partial torus. The radius of the latitude-circle of the partial torus ranges from R_{min} to R_{max}. Besides, the area ranging from the phantom circle to the largest circle is the projection of the partial sphere.

Again, let's analyze the known view of the intersection. If a capital letter P is assigned to the cutting-plane, P_W is its profile piercing line. Evidently, cutting-plane P is parallel to plane V. The left view shows that the partial torus and the partial sphere are cut off by cutting-plane P while the cylinder is not cut off as P_W does not contact the small circumference. The intersection of cutting-plane P and the partial sphere is a frontal latitude-circle while the intersection of cutting-plane P and the partial torus is a curve. Intersection of cutting-plane P and the composite revolution is a closed composite curved line in space as shown in Fig. 4.13 (a). The left view of the intersection is located on P_W. The front view of the intersection is required to be shown.

Graphic solution is as follows.

Step 1 Show the front view of the intersection on the spherical surface.

P_W intersects the large circumference at f'', which is the left view of the common point on the spherical surface. Through f'', draw its projection line to the front view to intersect the left-right symmetric plane of the sphere at f'. In the front view, with o' as center and $o'f'$ as radius, draw the frontal latitude-circle, which intersects the boundary of the partial sphere and the partial torus at a' and b'.

Step 2 Draw the front view of the intersection on the partial torus.

Projection points, a' and b', are the front views of the rightmost points (最右点) of the intersection on the partial torus. In order to find the front view of the leftmost point (最左点) of the intersection, a circle tangent to P_W is drawn in the left view. The circle is tangent to P_W at c'' and intersects the front-rear symmetric plane at g''. Through g'', draw its projection line to the front view to intersect the frontal outline at g'. From g', draw a line perpendicular to the axis of the composite revolution at c' which is the front view of the leftmost point on the intersection.

In order to find general common points, in the area belonging to the partial torus, take a proper radius from R_{min} to R_{max} as the radius and point o'' as center, and then draw a circle to intersect P_W at $1''$ and $2''$, and to intersect the front-rear symmetric plane at h''. Through h'', draw its projection line to the front view to intersect the frontal outline at h'. From h', draw a line perpendicular to the axis to intersect the projection line drawn from $1''$ at $1'$, and from $2''$ at $2'$ respectively.

Step 3 Connect points smoothly to obtain the view of the intersection.

In the front view, connect a', $2'$, c', $1'$, and b' smoothly with a thick line. Then curve $a'\,2'\,c'\,1'\,b'$ is the front view of the intersection on the partial torus.

Lastly, check, clean and brighten all the thick lines. So far, the front view of the intersection of the composite revolution has been completed.

4.3 Intersections of Two Revolutionary Surfaces (两回转体的交线)

We turn to study intersections of two revolutions in this section, which are actually the intersections of two revolutionary surfaces. The character of the intersections is similar to that of a plane and a revolution, i.e. they all consist of a series of common points—each point on the intersection is shared by both revolutionary surfaces.

In general, the intersection of two revolutionary surfaces is a spatial curve but in some special cases, the intersection may be a straight-line or a planar curve such as a circle or an ellipse.

Two methods are employed to find the views of the intersections of two revolutionary surfaces, which are finding-points-on-surface method (表面取点法) and cutting-plane method (切平面或辅助平面法).

4.3.1 Finding-points-on-surface method

Finding-points-on-surface method is a simple method to show the views of the intersection of two revolutionary surfaces. But it is only applicable when there is at least one cylindrical surface among the revolutionary surfaces. From 3.2.2.1, it is known that when a cylinder is projected along the cylinder's axis, the cylindrical surface appears as a circumference. Thus, the view of the intersection on the cylindrical surface is also on the circumference. Select enough common points on the circumference, find their other views by means of finding-points-on-surface method, and then connect all the projection points on the same view to form the view of the intersection.

Example 4.11 In Fig. 4.14, a frustum cone intersects a quarter of a cylinder. Complete the top view and the front view.

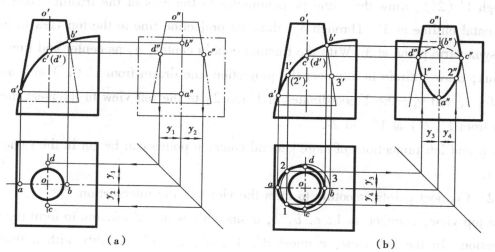

Fig. 4.14 Intersection on a cylinder and a cone.

Analysis is as follows.

First of all, draw the left views of the frustum cone and the quarter of a cylinder with phantom lines.

Three views of the center of the frustum cone are: o, o' o''.

Let's analyze the views of the intersection. The given views show that the axis of the frustum cone is perpendicular to plane H while the axis of the quarter of a cylinder is perpendicular to plane V. The intersection generated by the frustum cone intersecting the quarter of a cylinder is a closed spatial curve. The intersection appears as a circular arc on the circumference in the front view. However, its top and left views need to be found.

In the front view, two frontal outlines of the frustum cone intersect the quart of the circumference at a' and b', which are the front views of a pair of special common points on the intersection; the right-left symmetric plane of the frustum cone intersects the circumference at $c'(d')$, which is the front view of the other pair of special common points on the intersection. On the circumference between a' and b', proper projection points such as $1'(2')$, are picked up, which are the front views of a pair of general common points. Thus, to complete the top and left views of the intersection is actually to complete the top and left views of these common points.

Graphic solution is as follows.

Step 1 Complete the other views of the common points.

From a' and b', draw their projection lines to the top and left views respectively to intersect the front-rear symmetric plane at a and b in the top view and at a'' and (b'') in the left view.

From $c'(d')$, draw their projection line to left view to intersect the profile outlines of the frustum of the cone at c'' and d''.

Transfer y_1 and y_2—the Y coordinates of c'' and d'', to the top view to intersect the left-right symmetric plane of the frustum cone at c and d.

The top projection points, 1 and 2, may be obtained based on $1'(2')$ by means of latitude-circle because points I II are yet located on the conical surface.

Through $1'(2')$, draw their line perpendicular to the axis of the frustum cone to intersect the right frontal outline at $3'$. Through $3'$, draw its projection line to the top view to intersect the front-rear symmetric plane at 3. With the frustum cone's center, o, as center and line segment $o3$ as the radius, draw a circle to intersect the projection line drawn from $1'(2')$ at 1 and 2.

Transfer y_3 and y_4—the Y coordinates of 1 and 2, to the left view to intersect the projection line drawn from $1'(2')$ at $1''$ and $2''$.

The top and left projections of more general common points can be got in the same way (Fig. 4.14 (b)).

Step 2 Connect points smoothly to form the views of the intersection.

In the top view, connect a, 1, c, b, 2, a smoothly with a thick line to form the top view of the intersection. In the left view, connect d'', $1''$, a'', $2''$, c'' smoothly with a thick line and connect c'', (b''), d'' smoothly with a hidden line to form the left view of the intersection.

Step 3 Brighten the reserved profile outlines of the cone.

The front view shows that the existing profile outlines of the frustum cone are above $c'(d')$. In the left view, the frustum cone's profile outlines above c'' and d'' are brightened.

Finally, check, clean and brighten all the thick lines. Note that because the frustum cone is higher than the quarter of a cylinder, in the left view, the part hidden by the frustum cone is drawn with a hidden line.

Example 4.12 In Fig. 4.15(a), the axes of two cylinders with different diameters intersect perpendicularly. Complete the front view of the object.

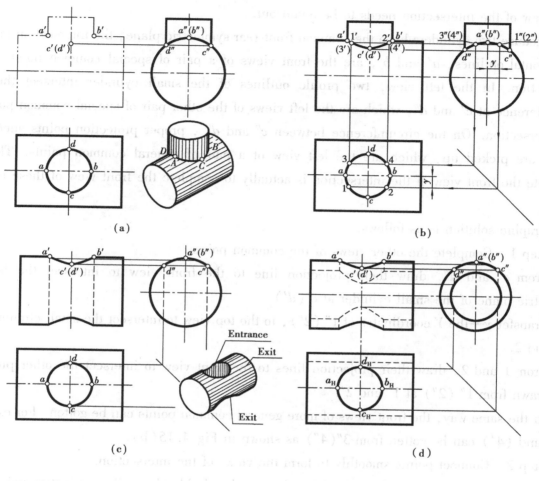

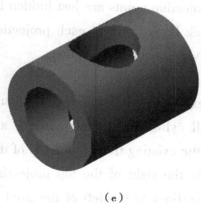

Fig. 4.15 Intersection on the surface of a cylinder.

Analysis is as follows.

First of all, draw the front view of the intersectional cylinders with phantom lines.

Let's analyze the given views and find out the known projections of the intersection.

97

The given views show that the axis of the large cylinder is perpendicular to plane W while the axis of the small cylinder is perpendicular to plane H. The intersection generated by two intersectional cylinders is a closed spatial curve line. The intersection appears as a small full circumference in the top view and as a large partial circumference in the left view. So, only the front view of the intersection needs to be found out.

Because two cylinders have the common front-rear symmetric plane, the intersection points of their frontal outlines, a' and b', are the front views of a pair of special common points on the intersection. In the left view, two profile outlines of the small cylinder intersect the large circumference at c'' and d'', which are the left views of the other pair of special common points on the intersection. On the circumference between c'' and d'', proper projection points such as $1''$ ($2''$), are picked up, which are the left view of a pair of general common points. Thus, to complete the front view of the intersection is actually to complete the front view of these common points.

Graphic solution is as follows.

Step 1 Complete the other views of the common points.

From c'' and d'', draw their projection line to the front view to intersect the left-right symmetric plane of the small cylinder at $c'(d')$.

Transfer y—the Y coordinates of $1''$ ($2''$), to the top view to intersect the small circumference at 1 and 2.

From 1 and 2, draw their projection lines to the front view to intersect the other projection line drawn from $1''$ ($2''$) at $1'$ and $2'$.

In the same way, the front views of more general common points can be gotten. For example, ($3'$) and ($4'$) can be gotten from $3''(4'')$ as shown in Fig. 4.15(b).

Step 2 Connect points smoothly to form the views of the intersection.

In the front view, invisible projection points are just hidden by visible projection points, so a hidden line is just hidden by a thick line. Connect each projection points from a' to b' smoothly with a thick line to form the front view of the intersection.

Step 3 Brighten the reserved outlines.

The left view shows that the existing frontal outlines of the small cylinder are above a'' (b''). Thus, in the front view, the small cylinder's frontal outlines above a' and b' are brightened. Similarly, the top view shows that the existing frontal outlines of the large cylinder are to the left of the top projection point, a, and to the right of the top projection point, b. Thus, in the front view, the large cylinder's frontal outlines to the left of the front projection point, a', and to the right of the front projection point, b', are brightened.

Lastly, check, clean and brighten all the thick lines as shown Fig. 4.15(b).

Fig. 4.15(c) is similar to Fig 4.15(a). The difference between them is that the small cylinder is replaced by a hole of cylindrical surface. The frontal outlines and profile outlines of the hole are invisible and they are drawn with hidden lines. Evidently, there are two intersections on the entrance (入口) and the exit (出口) of the hole. The two intersections appears as small full circumference in the top view and as two circular arcs on the large circumference between two hidden lines called profile outlines. Only the front views of the two intersections need to be found. Procedure to find out the front view of the two intersections is the same as that aforementioned in Fig. 4.15(a).

Fig. 4.15 (d) is a cylindrical sleeve with a vertical hole of cylindrical surface. The frontal outlines and horizontal outlines of the interior cylindrical surface in the cylindrical sleeve are invisible and they are drawn with hidden lines. Similarly, the frontal outlines and profile outlines of the vertical hole are also invisible and they are drawn with hidden lines.

Evidently, the vertical hole's surface intersects the outside cylindrical surface of the cylinder sleeve to generate two outside intersections. However, the vertical hole's surface intersects the inside cylindrical surface of the cylinder sleeve to generate two inside intersections as shown in Fig. 4.15(d). Thus, there are four intersections on the object. The two outside intersections are the same with that in Fig. 4.15(c). While the two inside intersections are similar to that in Fig. 4.15(c). The difference between them is that the inside intersections are drawn with hidden lines as shown in the front view in Fig. 4.15(d).

4.3.2 Intersection of surface — cutting-plane method

The cutting-plane method is an efficient and useful method for finding intersections' views.

As mentioned above, the key point to make the intersection's views is to find out the common points. Suppose that two intersectional revolutions are cut by cutting-plane P within intersection range, the cutting-plane P intersects two revolutions respectively to generate intersections which intersect each other to generate common points.

In Fig. 4.16, an object is formed by a frustum cone intersecting a cylinder. Suppose that when the object is cut by cutting-plane P, intersections are generated. The intersection of the cutting-plane P and the frustum cone is a latitude-circle while the intersections of the cutting-plane P and the cylinder are two straight-element lines L_1 and L_2. These intersections intersect mutually at points A, B, C, D which are common points on the intersection of the cylinder and frustum cone.

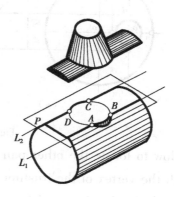

Fig. 4.16 Cutting-plane method.

In the same way, a series of common points may be obtained. These common points are connected smoothly to form the intersection's views.

The principle to select cutting planes is as follows.

1) Cutting-plane should be within intersection range.

2) When the cutting-plane intersects two revolutionary surfaces, the intersection must be either a straight line or a circle.

Example 4.13 As shown in Fig. 4.17, an object is formed by a frustum cone intersecting a hemisphere. Complete the views of the intersection.

Analysis is as follows.

First of all, draw the left view of the two intersectional revolutions with phantom lines. The three views of the hemisphere's center are o', o, o'' and that of the frustum cone's center are o_1', o_1, o_1''.

The given views show that the upper hemisphere intersects a frustum cone to generate a intersection which is a closed spatial curve line. Since there is no cylinder in two intersectional revolutions and, any of the three views of the intersection is unknown, finding-points-on-surface method is invalid. The cutting-plane method must be used to find three views of the intersection.

Graphic solution is as follows.

Step 1 Locat the special common points.

Because the two intersectional revolutions have the common front-rear symmetric plane, the intersection points of their frontal outlines, a' and b', are the front view of a pair of special common points. From a' and b', draw their projection lines to the top and left views to intersect the front-rear symmetric plane at a and b in the top view, and at a'' and (b'') in the left view.

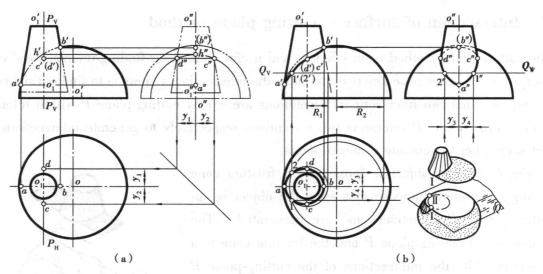

Fig. 4.17 Intersection of a sphere and a cone.

How to locate the other pair of special common points? Suppose that cutting-plane P passes through the vertex of the frustum cone and it is parallel to plane W, so P_V and P_H are respectively its frontal piercing line and horizontal piercing line. The two intersectional revolutions are cut by the cutting-plane P to generate intersections. The intersections of the cutting-plane P and the frustum cone are profile outlines of the frustum cone. Simultaneously, the intersection of the cutting-plane P and the hemisphere is a profile semicircle which intersects the profile outlines of the frustum cone to generate common points. Drawing procedures are as follows.

P_V intersects the frontal outlines of the hemisphere at h'. From h', draw its projection line to the left view to intersect the front-rear symmetric plane at h''.

With $o''h''$ as radius and the hemisphere center o'' as center, draw a semicircle which intersects the two profile outlines of the frustum cone at c'' and d'''.

Through c'' and d''', draw their projection line to the front view to intersect P_V at c' (d').

Transfer y_1 and y_2—the Y coordinates of c'' and d''', to the top view to intersect P_H at c and d.

Step 2 Locat general common points.

Suppose that cutting-plane Q is perpendicular to the axis of the frustum cone, or parallel to plane H, so Q_V and Q_W are respectively its frontal piercing line and profile piercing line. The two intersectional revolutions are cut by cutting-plane Q to generate intersections. The intersection of the cutting-plane Q and the frustum cone is a small horizontal latitude-circle. Simultaneously, the intersection of the cutting-plane Q and the hemisphere is a big horizontal latitude-circle which intersects the small horizontal latitude-circle to generate common points. Drawing procedures are as follows.

Q_V intersects the frustum cone to obtain the radius R_1 and intersects the hemisphere to obtain the radius R_2 (Fig. 4.17(b)).

In the top view, with R_1 as radius and the frustum cone's center o_1 as the center, draw a small horizontal latitude-circle. With R_2 as radius and the hemisphere's center o as the center, draw a big horizontal latitude-circle. The two circles intersect each other at 1 and 2.

Through 1, 2, draw their projection line to the front view to intersect Q_V at $1'$ $(2')$.

Transfer y_3 and y_4—the Y coordinates of the 1 and 2, to the left view to intersect Q_W at $1''$ and $2''$.

In the same way, the three views of more general points can be got.

Step 3 Connect points smoothly to form the views of the intersections.

In the top view, connect a, 1, c, b, 2, d, a with a thick line smoothly. In the left view, connect d'', $2''$, a'', $1''$, c'' smoothly with a thick line and connect c'', (b''), d'' smoothly with a hidden line.

Step 4 Brighten the reserved outlines.

The front view shows that the existing profile outlines of the frustum cone are above c' (d'). In the left view, the frustum cone's profile outlines above c'' and d'' are brightened.

Lastly, check, clean and brighten all the thick lines. Note that as the frustum cone is higher than the hemisphere, the hemisphere's profile outline hidden by the frustum cone is drawn with a hidden line.

4.3.3 Special case of intersection

4.3.3.1 Intersections—ellipses

The sphere enveloped (包裹) by the two cylinders is tangent to the surfaces of both cylinders only when they have equal diameters. When two cylinders with the same diameter intersect, the intersections are changed from spatial curves to plane curves—two elliptical curves (Figs. 4.18 (a) to (b)). The two elliptical curves appear as a circumference representing a vertical cylinder in the top view and as a circumference representing a horizontal cylinder in the left view.

However, they appear as two straight line segments in the front view (Fig. 4.18 (c)). How to draw the straight line segments? As shown in Fig. 4.18 (c), the frontal outlines of the vertical cylinder and the horizontal cylinder intersect at a', c', b', and d', which are the front views of the special common points on the ellipses. Two common points located diagonal (对角线) are connected to form one straight line, e.g. line segment $a'b'$. The other two common points located diagonal are also connected to form the other straight line, e.g. line segment $c'd'$.

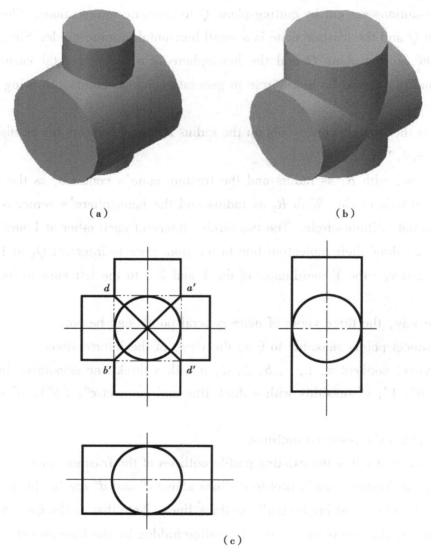

Fig. 4.18 Intersections—ellipses.

4.3.3.2 Intersections—circles

When two or more intersectional revolutions share one axis (同轴相贯), their intersections are circles perpendicular to their axes—latitude-circles (Fig. 4.19).

4.3.3.3 Intersections—straight lines

In Fig. 4.20(a), when two cylinders with parallel axes intersect, the intersections are straight lines, two elements (两素线). While two cones with a common vertex intersect, the intersections are also straight lines, two elements (Fig. 4.20(b)).

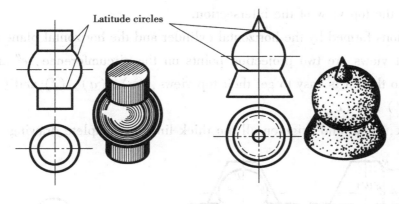

Fig. 4.19 Intersection—circle

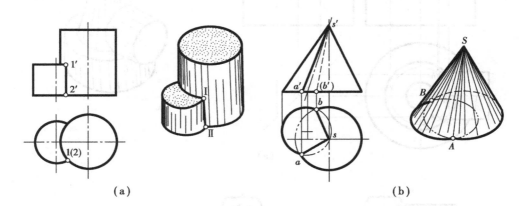

(a)   (b)

Fig. 4.20 Intersections—straight lines

4.3.4 Intersections of multiple objects (多体相贯)

It is called multiple objects intersecting in that three or more than three primary revolutions intersect together. In fact, the intersections of multiple objects are composed of each of the segment intersection which is the result of every two adjacent revolutionary surfaces intersecting. The steps to find the views of the intersections are as follows:

1) Find out dividing points of each segment intersection.

2) Draw the views of the intersections of every two adjacent intersecting objects in turn with the methods discussed above.

Example 4.14 Fig. 4.21 shows the object composed of a vertical cylinder, a horizontal cylinder and a frustum cone. Complete the front and top views of the object.

As shown in Fig. 4.21(b), the horizontal cylinder intersects both the frustum cone and the vertical cylinder. Each segment intersection appears as a partial circumference representing the horizontal cylinder in the left view.

In Fig. 4.21(a), arc $a''g''b''h''c''$ is the left view of the intersection formed by the horizontal cylinder intersecting the frustum cone. The method mentioned above can be used to find the front view and the top view of the intersection.

In Fig. 4.21(c), arc $e''d''f''$ is the left view of the intersection formed by the horizontal cylinder intersecting the vertical cylinder. The method mentioned above can also be used to find the front view and the top view of the intersection.

The intersections formed by the horizontal cylinder and the horizontal plane Q are two straight lines and their left views are two projection points on the circumference, e'' and (a''), f'' and (c''). According to then it is easy to get their top views, (e), (a), (f) and (c) and their front views, e', a', (f') and (c').

Lastly, check, clean and brighten all the thick lines to complete drawings.

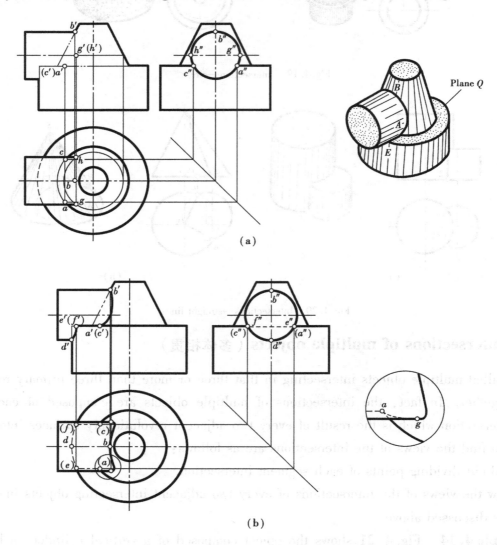

Fig. 4.21 Intersections of multi-object.

Chapter 5 Composite Objects

The previous study involves in the representation of a point, a line, a plane and some primary objects such as prisms, cylinders and so on. Our purpose in this chapter is to study the representation of a composite object which is composed of several primary objects. Three aspects including drawing views, dimensioning and reading views will be introduced in this chapter.

5.1 Projection Rules of an Object

5.1.1 Three views

As discussed in previous chapters, suppose that an object is put properly, it is projected towards three projection planes to form front view, top view and left view (Fig. 5.1 (a)). After the object is removed, plane V is imagined to be fixed and plane H as well as plane W to be hinged (铰接) with plane V and they are unfolded outwards to lay the three views on drawing paper in Fig. 5.1 (b).

From now on, three projection axes, even 45° miter-line are not drawn in engineering drawings, but they exist actually, so the relationship that projection lines are perpendicular to projection axes is yet kept in the adjacent views as shown in Fig. 5.1 (b). When projection axes are hidden, the frontal surface or the rear surface of the object is taken as reference line to measure depth dimensions between the top view and the left view. If the depth between the left view and the top view needs to be transferred, we use divider. Furthermore, the distance between the top view and the front view, and that between the left view and the front view are not necessarily equal, but the depth dimensions between the top view and the left view should be the same.

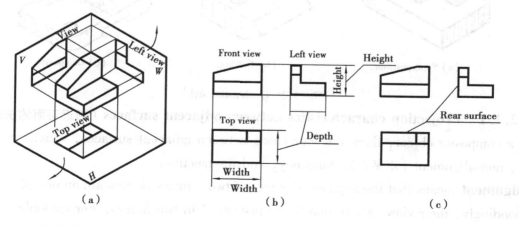

Fig. 5.1 The three views.

5.1.2 Projection rules of an object（物体的投影规律）

Based on the forming of the three views of an object, the projection rules of an object are stated as follows (Fig. 5.1 (b)):

The front and top views are aligned vertically to show the width of the object（长对正）.

The front and left views are aligned horizontally to show the height of the object（高平齐）.

The top and left views have the same depth of the object（宽相等）.

This projection rules indicate the arrangement of the three views, which is also stipulated by the Chinese National Standard of Technical Drawing. To draw a view without following the arrangement is a serious error and is generally regarded as one of the worst mistakes possible in drawing. In the three views, the name of each view is omitted as shown in Fig. 5.1 (c).

Furthermore, it is very important to distinguish the frontal surface and the rear surface in the views. In the left and top views, the two rear surfaces face towards the front view, respectively.

5.2 Drawing Three Views

5.2.1 Analysis

5.2.1.1 Form of a composite object

Any composite object can be broken down into a combination of some primary geometric objects. Any of these basic shapes can be positive, the superposition style（叠加 Fig. 5.2 (a)) or negtive, the cutting style（切割 Fig. 5.2 (b)). For example, a hole is a negtive cylinder. Usually, the two combination styles metioned above can be used to form a composite object as shown in Fig. 5.2 (c).

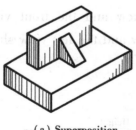

(a) Superposition.　　　(b) Cutting.　　　(c) Composite.

Fig. 5.2　Combination style.

5.2.1.2 Projection characteristics between adjacent surfaces（邻近表面的投影特征）

In a composite object, there are four cases between adjacent surfaces, which are alignment（平齐）, non-alignment（不平齐）, tangency, and intersection.

Alignment means that the adjacent surfaces of two primary objects are on one plane actually. Correspondingly, their views are in one frame instead of in two frames. For example, in Fig 5.3 (a), surface A aligns to surface B, so figure (b) is correct and figure (c) is incorrect.

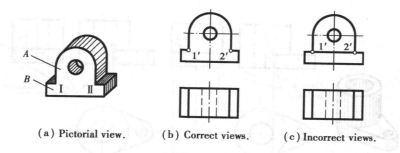

(a) Pictorial view.　　(b) Correct views.　　(c) Incorrect views.

Fig. 5.3　The view of alignment.

Non-alignment is contrary to alignment, that is to say, adjacent surfaces of two primary objects are on two different planes. Correspondingly, their views are in two frames instead of in one frame. For example, in Fig. 5.4(a), surfaces A and B are on different planes; thus, figure (b) is correct and figure (c) is incorrect.

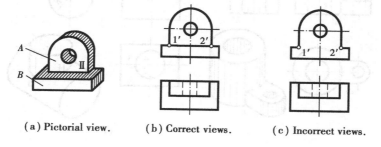

(a) Pictorial view.　　(b) Correct views.　　(c) Incorrect views.

Fig. 5.4　The view of non-alignment.

Tangency means that the adjacent surfaces of two objects are one continuous plane. Correspondingly, its view is one frame but not two frames. In Fig. 5.5, the front and rear surfaces of the motherboard (底板) are tangent to the external surface of the vertical cylinder; thus, There are no lines at tangencies either in the front view or in the left view. In other words, both front and left views are one frame. So, in Fig. 5.5, figure (b) is correct and figure (c) is incorrect. As explained in Fig. 4.12, since the partial torus is tangent to a cylinder at the left end and tangent to a partial sphere at the right end, no border lines are located at tangencies.

Intersection is contrary to tangency. When a plane surface intersects a curved surface or both two curved surfaces intersect, intersections are generated. In Fig. 5.6 (a), the front and rear surfaces of the motherboard intersect the external surface of the vertical cylinder; thus, both in the front view or in the left view, there are border lines in the adjacent frames. In Fig. 5.6 (b), a small cylinder intersects a big cylinder to generate a intersection. The front view of the intersection may be obtained by the finding-points-on-surface method, as explained in 4.1.1, and it may be obtained approximately by drawing an arc whose radius is the radius R of the big cylinder. Note that when two axes of two cylinder intersect perpendicularly, the intersection protrudes towards (凸向) the axis of the big cylinder. Drawing method of the intersection is as follows. With common point $1'$ as center and R as radius, draw an arc to intersect the axis of the small cylinder at O_0 first and then with O_0 as center and R as radius, draw arc $1'2'3'$ which is the front view of the approximate intersection.

Fundamentals of Engineering Drawing

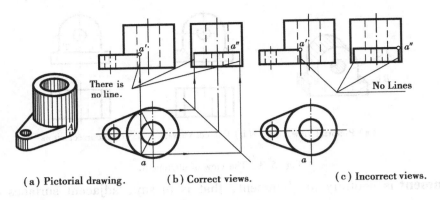

(a) Pictorial drawing.   (b) Correct views.   (c) Incorrect views.

Fig. 5.5 Projection characteristic of tangency.

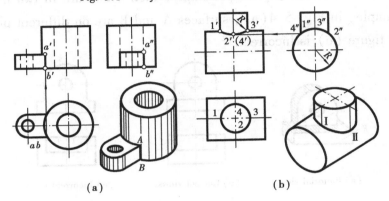

(a)   (b)

Fig. 5.6 Projection characteristic of intersection.

5.2.2 Drawing three views of a composite object（组合体三视图的绘制）

5.2.2.1 Analyzing-shape method（形体分析法）

This chapter involves in three aspects including drawing views, dimensioning and reading views. As a useful method, analyzing-shape method, is introduced to study the three aspects. Basic idea of this method is to suppose that a composite object can be disassembled（分解）into some primary objects first and then analyze their relative positions, combined styles as well as the cases of adjacent surfaces. In other words, the problem to deal with a composite object is transformed（转换）into the problem to deal with some primary objects one by one. For example, in Fig. 5.7(a), the composite object may be disassembled into a vertical cylinder sleeve（套筒, 圆筒）, a motherboard（底板）, a rib（肋板）, and a horizontal cylinder sleeve. The four components are superposed together actually. Holes of the motherboard, the vertical and horizontal cylinder sleeves are formed by cutting. In superposition, intersections are generated. The front and rear surfaces of the rib intersect the outside cylindrical surface of the vertical cylinder sleeve to generate two intersections—two elements. The outside cylindrical surface of the vertical cylinder sleeve intersects the outside of the horizontal cylinder sleeves to generate an external intersection（外交线）. Similarly, two inside cylindrical surfaces of the two cylinder sleeves intersect to generate an internal intersection. However, the front and rear surfaces of the motherboard are tangent to the vertical cylinder sleeve, respectively.

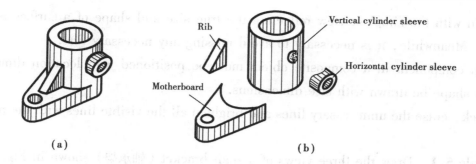

Fig. 5.7 A composite object.

5.2.2.2 Select the projection direction for the front view

Because the front view is the most important one in the three views, it is very important to select adequate projection direction to form the front view. Good projection directions feature as follows.

1) It makes the front view show shape feature of the composite object mostly.

2) It makes the front view express as many true surfaces as possible. Namely, let a composite object be placed in a position where its major surfaces are parallel to plane V. If a composite object is a revolution, let its axis be perpendicular to plane W.

3) It makes as few hidden lines in the front views as possible.

5.2.2.3 Determining adequate scale and drawing paper size

Based on the size and complexity of the composite object, proper scale and drawing paper are determined. The original value scale 1∶1 should be considered first. In the case of a small composite object, the ratio of enlargement scale such as 2∶1 or 5∶1 should be applied. On the contrary, the ratio of reduction scale such as 1∶2 or 1∶5 is adopted. See Chart 1.2 Scales. Drawing paper is selected according to Chart 1.1 Paper Size.

To give a well-balanced appearance of the three views in a sheet of paper, each view should be arranged evenly. As shown in Fig. 5.8, Capital letter A is assigned to the horizontal spacing (水平间隔) and capital letter C to the vertical spacing (垂直间隔). Small letters l, t, h stand for the width (长), depth (宽) and height (高) of the object respectively; L and B represent the width and height of the drawing paper respeclively. If scale 1∶1 is adopted, the formula for calculating the

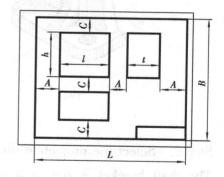

Fig. 5.8 Arrangement the three views.

horizontal spacing and the vertical spacing is as follows: $A = (L - l - t)/3$, and $C = (B - h - t)/3$.

5.2.2.4 Drawing the three views

First of all, draw drafting drawings (底图) in a very light stroke with a hard pencil of H or HB. Several points should be noticed in this process.

1) When drawing the three views of a composite object, start from the largest component (primary object) of it; when drawing any to component (primary object), start from its external structure to its internal structure.

2) Begin with the simplest view reflecting the true size and shape of a surface and then the other views. Meanwhile, it is necessary to avoid missing any necessary line.

3) Each component in a composite object must be positioned with location dimensions first and then its shape be drawn with size dimensions.

4) Check, erase the unnecessary lines and brighten all the visible lines with the proper pencil of B or 2B.

Example 5.1 Draw the three views of a shaft bracket (轴承架) shown in Fig. 5.9 (a).

Step 1 Analyze the shape.

Fig. 5.9 (b) shows that the shaft bracket can be disassembled into four components, motherboard, holder plate (支承板), rib, and cylinder sleeve. Actually, the four components are superposed together. Two holes of the motherboard are negative cylinders. Similarly, a small hole at the top of the cylinder sleeve is a negative cylinder. The intersections generated in superposition are as follows. The left and right surfaces of the rib intersect the outside cylindrical surface of the cylinder sleeve to generate two intersections—two elements. The small hole at the top of the cylinder sleeve intersects the outside cylindrical surface of the cylinder sleeve to generate an external intersection (外交线). Similarly, the small hole intersects the inside cylindrical surface of the cylinder sleeve to generate an internal intersection (内交线). However, the left and right surfaces of the holder plate are tangent to the cylinder sleeve, respectively.

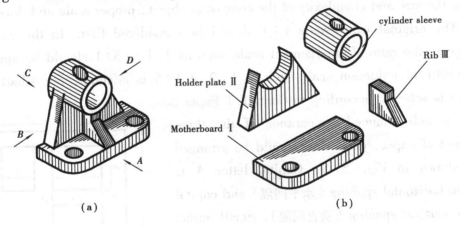

Fig. 5.9 Shaft bracket.

Step 2 Select the projection direction for the front view.

The shaft bracket is put in a conventional position as shown in Fig. 5.9(a). A good projection direction is chosen from the arrow A, B, C on D. The projection direction of the arrow A has obvious advantages if it is compared with the others. Namely, the front view obtained based on the projection direction of the arrow A shows the shape feature mostly and fewer hidden lines.

Step 3 Choose the scale and a sheet of drawing paper.

Based on drawing-paper's L and B as well as l, t, h of the object as shown in Fig. 5.8, horizontal spacing, A, and vertical spacing, B, can be calculated. Fwthermore, values a, b, c and d are calculated as follows.

$a = A + l/2$. It means that a equals the summation of the value of the horizontal spacing and half of the width of the object.

$b = C + h$. It means that b equals the summation of the value of the vertical spacing and the height of the object.

$c = C + t$. It means that c equals the summation of the value of the vertical spacing and the depth of the object.

$d = A + t$. It means that d equals the summation of the value of the horizontal spacing and the depth of the object.

Based on the frame line (图框线) of the drawing-paper, the basic lines which are the object's base, object's left-right symmetrical plane as well as the object's rear surface can be determined by means of values a, b, c, and d as shown in Fig. 5.10 (a).

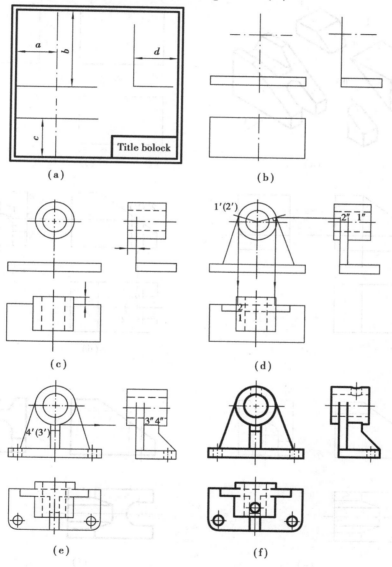

Fig. 5.10 Steps of drawing three views of the shaft bracket.

Step 4 Draw three views (Figs. 5.10 (b) to (e)).

1) Draw three views of the motherboard as shown in figure (b).
2) The cylinder sleeve's three views are drawn in figure (c).

3) The holder plate is represented in figure (d). Note that it is important to find out the correct tangent points.

4) Figure (e) expresses the rib. It is important to draw the intersections generated by the left and right surfaces of the rib intersecting the outside cylindrical surface of the cylinder sleeve.

5) The small hole is drawn in figure (f). Note that the outside and inside intersections are projected correctly in the left view.

Lastly, check, erase unnecessary lines and brighten all the visible lines with proper pencil of B or 2B as shown in Fig. 5.10 (f).

Example 5.2 Draw three views of a composite object as shown in Fig. 5.11(a)

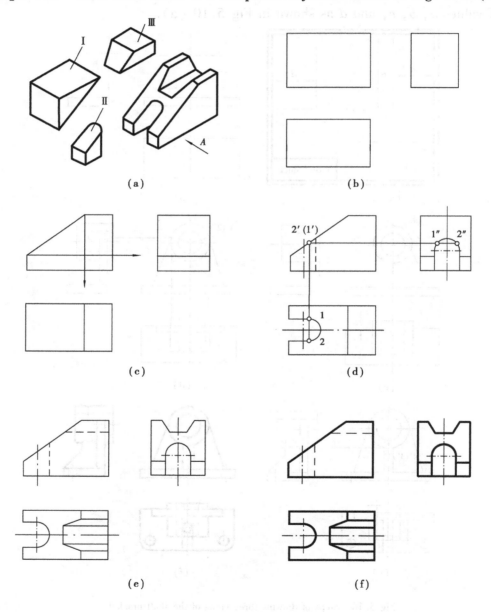

Fig. 5.11 Steps of drawing three views of a cutting object.

Step 1 Analyze the shape.

As shown in Fig. 5.11(a), the composite object is formed by cutting off parts I, II, III from a rectangular prism.

Step 2 Select the projection direction for the front view.

Projection direction of the arrow A is chosen to form the front view (Fig. 5.11(a)).

Step 3 Choose the scale as well as a sheet of drawingpaper and arrange the three views evenly in the drawing paper.

Step 4 Draw three views.

Draw the complete rectangular prism first and then cut off parts Ⅰ, Ⅱ, Ⅲ one by one. See Figs. 5.11(b) to (e).

Part Ⅰ is cut off, where a plane perpendicular to plane V is left. The plane appears as an edge in the front view and as foreshortened surfaces in the top and left views, respectively. Usually, among the three views, the edge view is drawn first. Namely, in Fig. 5.11(c), the front view of the plane is drawn first and then the top and left views of the plane.

Part Ⅱ is cut off, where a U-type slot is left. The slot consists of three surfaces which are two frontal planes and a half cylindrical surface. The top view of the slot is drawn first and then the front view. In addition, its left view is drawn as shown in figure (d).

Part Ⅲ is cut off, where a swallow-tailed slot (燕尾状的槽) is left. The slot consists of three surfaces which are two slot side surfaces (槽壁面) perpendicular to plane W and one slot bottom surface (槽底面) parallel to plane H. First, the left view of the slot is drawn. Secondly, the front view of it is drawn. Lastly, the top view of it is drawn as shown in figure (e).

Finally, check, erase unnecessary lines and brighten all the visible lines with the proper pencil of B or 2B. See Fig. 5.11(f).

5.3 Dimensioning

5.3.1 Requirements of dimensioning

The three views are usually used to describe the shape of an object. However, an engineering drawing must have a complete size description. In dimensioning, three vital rules will be followed.

1. Each feature is dimensioned and positioned only once.

2. Each feature is dimensioned and positioned where its shape shows.

3. The dimensioning must accord with the regulation of the Chinese National Standard for Technical Drawing. These rules help to clarify the reading of any drawing and reduce the chance of error and confusion for the person making the part.

5.3.2 Dimensioning primary object (基本立体的尺寸注法)

5.3.2.1 Dimensioning polyhedra

Usually, the three basic dimensions, width (长), depth (宽) and height (高), determine the size of a primary object. Therefore, the size dimensions (形状尺寸) of a primary object include dimensions in width, depth and height.

For a polyhedron such as a prism or a pyramid, their size dimensions are shown in Fig. 5.12. The width and depth are given in the top view and the height in the front view. Note that the dimension (16.2) is a reference dimension (参考尺寸) of the hexagon while the dimension 8×8 and 12×12 indicate that the side length (边长) of the two square bases is 8 and 12 respectively.

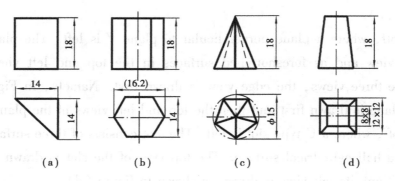

Fig. 5.12 Dimensioning for polyhedra.

5.3.2.2 Dimensioning revolutions

The general method of dimensioning a cylinder is to give both its diameter and its altitude in the rectangular view as shown in Fig. 5.13 (a). Note that when we are dimensioning diameter, symbol "ϕ" leads the number of diametral dimension. Again, the radius of a cylinder should never be given because measuring tools, such as the micrometer (千分尺) and caliper (卡钳), are designed to check diameters. In some case, the symbol ϕ can help eliminate the view of circle. For example, a cone is dimensioned by giving its diameter of its base and its altitude in the triangular view. A frustum of a cone may be dimensioned by giving the diameters of its two bases and its altitude in the isosceles trapezoid (等腰梯形) view (Fig. 5.13 (b)). In Fig. 5.13 (c), a torus is dimensioned by giving two diameters, in which one is the diameter of the generatrix circle, $\phi 8$, and the other is the diameter of the locus circle of the circular center of the generatrix circle (母线圆的圆心轨迹圆), $\phi 16$. A sphere only requires the diameter of the sphere. The spherical diameter symbol "$S\phi$" leads the number of the spherical diametral dimension (Fig. 5.13(d)).

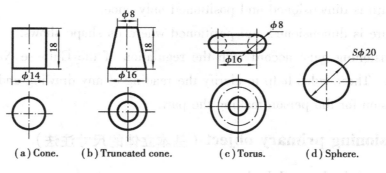

(a) Cone.　　(b) Truncated cone.　　(c) Torus.　　(d) Sphere.

Fig. 5.13 Dimensioning for revolutions.

5.3.2.3 Dimensioning objects with slot

When a small part is cut off from a large object, a slot is formed. Location dimensions (位置尺寸) are used to determine the positions of these cutting-planes.

An object with a slot is dimensioned by giving its size dimensions first and then location dimensions. Fig. 5.14(a) shows an object with a slot whose location dimensions are 6 and 8. It is unnecessary to give the dimension of the intersection because as long as the cutting-planes are located, the intersections are determined. Similarly, in the Fig. 5.14, the location dimensions in figure (b) are also 6 and 8. In figure (c), location dimensions are 5 and 13. In figure (d), location dimensions are both 5 and 5.

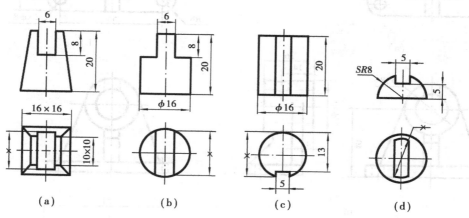

Fig. 5.14 Dimension for uncompleted basic objects.

5.3.3 Dimensioning a composite object

5.3.3.1 Dimension datums (尺寸基准):

Datum refers to a point, a line, or a surface used as a reference in surveying (测量), mapping, or geology. Three dimension datums are required in a composite object, which are dimension datums in width, in depth and in height. Generally, in a composite object, the symmetrical plane, bottom surface or the revolution's axis is selected as dimension datum in width, depth and height, respectively.

5.3.3.2 Analyzing the shape

It is much more complicated to dimension a composite object than to dimension a principal object. However, a composite object can be regarded as a combination of primary objects, which makes dimensioning easier. In dimensioning a composite object, general dimensions (总体尺寸) are another kind of dimensions besides size dimensions and location dimensions. General dimensions of a composite object include the total width, total depth and total height.

5.3.3.3 Dimensioning procedure

The steps of dimensioning a composite object are shown in Example 5.3.

Example 5.3 Dimension the shaft bracket (Fig. 5.15(a)).

Step 1 Analyze the shape.

As stated previously, the shaft bracket is imagined to be disassembled into four

Fundamentals of Engineering Drawing

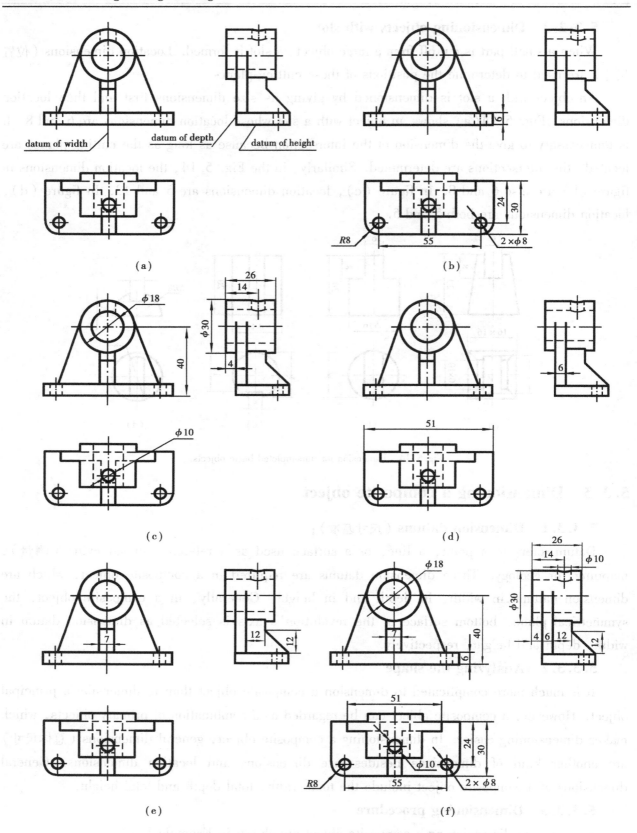

Fig. 5.15 Dimensioning the shaft bracket.

components—motherboard, cylinder sleeve, holder plate and rib as shown in Fig. 5.9.

Step 2 Determine dimension datums.

The shaft bracket's left-right symmetric plane, the bottom surface as well as the back surface of the holder plate in the shaft bracket are selected as the datums in width, height and thickness (depth) as shown in Fig. 5.15(a)

Step3 Dimension the location and size dimensions for each component.

The motherboard is dimensioned by giving its location dimensions, 55 and 24, and its size dimensions, $R8$, $2\times\phi 8$, 6, as shown in Fig. 5.15 (b). The cylinder sleeve is dimensioned by giving its location dimensions, 40 and 4, and its size dimensions, $\phi 30$, $\phi 18$, 26, as shown in Fig. 5.15(c). The small hole at the top of the cylinder sleeve is dimensioned by giving its location dimension 14 and its size dimension $\phi 10$, as shown in Fig. 5.15(c). The holder plate is dimensioned by giving its location dimensions, 6 and 51, and its size dimensions, 6 and 51, as shown in Fig. 5.15 (d). It should be pointed out that in some cases, one dimension may have two functions——determining location and size, for example the dimensions 6 and 51. The rib is dimensioned by giving its location dimensions, 7, 12 and 12, as shown in Fig. 5.15(e). Note that when the position of the rib is located, the size of it is determined. Therefore, the size dimensions are the same with the location dimensions.

Step 4 Give general dimensions.

The general dimension of the shaft bracket in width is that of the motherboard. That is to say, it is the summation of the location dimension 55 and double of the size dimension $R8$. General dimension in depth is the summation of three dimensions, $(24 + 8 + 4)$. General dimension in height is the summation of the location dimension 40 and the size dimension $\phi 30/2$. Note that when a general dimension is represented by the summation of location dimensions and size dimensions, it is generally unnecessary to be labeled. For example, the total width of the shaft bracket, 71, is not dimensioned directly. Similarly, the general dimensions in depth and height are also not dimensioned directly.

Lastly, it is necessary to check and complete dimensioning (Fig. 5.15(f)).

5.3.3.4 Several notices

It must be emphasized again that dimensioning must obey the stipulation of the Chinese National Standard for Technical Drawing.

Dimensions are usually placed outside the views. But if dimension extensions traverse many lines, the dimension should be placed inside the views. In Fig. 5.15 (f), dimensions 4, 6, 12 are placed inside the left view.

Dimensions should be placed in the view that shows shape feature mostly, and the dimensions of the same direction should be given in one view as possible as you can to make it convenient to check. For example, Fig. 5.16 (a) shows the correct dimensioning because all the dimensions of the left end of the object are dimensioned in the front view while the dimensions of the right end of the object are dimensioned in the top view. However, Fig. 5.16 (b) is the incorrect dimensioning. In addition, dimensions shouldn't be dimensioned on hidden lines.

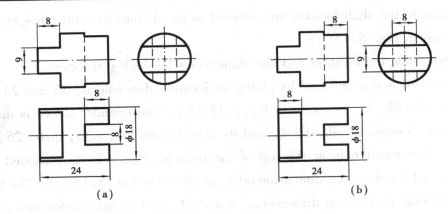

Fig. 5.16 Notices in dimensioning.

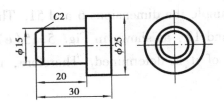

Fig. 5.17 Dimensioning a chamfer.

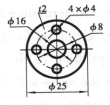

Fig. 5.18 Dimensioning a thin plate.

A chamfer (倒角) is a frustum cone with a vertex of 90° vertex angle is 90° generally. If a chamfer is dimensioned by giving $C2$ on a leader of a beveled edge, "C" means chamfer and "2" means the axial dimension (轴向尺寸) of the frustum of a cone.

Dimensions of a thin plate (薄板) are shown in Fig. 5.18. "$t2$" denotes two meanings. Letter t is the abbreviation (缩写) of thickness while 2 means that thickness of the thin plate is 2 mm.

Figs. 5.19 (a) to (b) are two examples of dimensioning composite objects. They are analyzed by readers to find out dimension datums in height, width and depth, location dimensions and size dimensions.

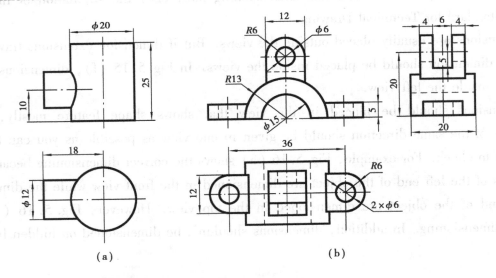

Fig. 5.19 Practice.

5.4 Reading Views(读图)

Based on the given views and according to the projection rules, an object can be conceived and this procedure is called "reading views". Completing the third view with the help of the given views is a good way to test whether students really understand the views and figure out the object. Drawing views and reading views are two important skills for students in this course. The former aims at representing an object with views and the later aims at visualizing(想象) an object shown by views.

5.4.1 Basic knowledge for reading views

Generally, one view cannot express the object, so when reading views, at least two views should be read simultaneously. As shown in Fig. 5.20, all the front views are the same, however they are different objects since their top views are different.

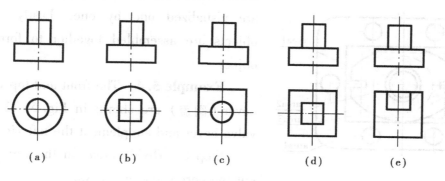

Fig. 5.20 Different objects.

In order to read views effectively, students must be familiar with the three views of primary objects first. As mentioned before, a composite object can be imagined to be disassembled into some primary objects. Therefore, the three views of primary objects such as cones, cylinders, spheres, prisms and pyramids are not only the fundamental in drawing views, but also the fundamental in reading views.

It is important to understand the meaning of each line in the views. Any thick line or hidden line in the views may represent a surface or an outline of a revolutionary surface or an intersection of two surfaces. Any closed area can be regarded as the projection of a primary object or a closed planar surface or a curved surface or a through hole and so on. As shown in Fig. 5.21, line A is the projection of top planar surface; line B is the projection of the outline of the cylindrical surface; line C is the intersection of two surfaces. The

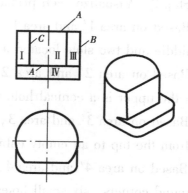

Fig. 5.21 Meanings of lines and areas in view.

closed areas Ⅰ and Ⅲ are the projection of the curved surfaces; the closed area Ⅱ is the projection of the planar surface; also, the closed area Ⅳ is the projection of a composite surface combined by a plane and two curved surfaces.

5.4.2 Methods to read views

5.4.2.1 Analyzing-shape method

As discussed before, analyzing shape method is important for us to study this chapter and it may be applied to draw views, read views and dimension. However, how to apply the analyzing-shape method in reading views? Its fundamental idea is that an area in a view represents a primary object in space. Therefore, any view is divided into many areas first and then the corresponding projection of each area is found out in other views according to the projection rules. Whereafter, the primary objects shown by views are visualized one by one. Lastly, these primary objects are assembled together to form a composite object.

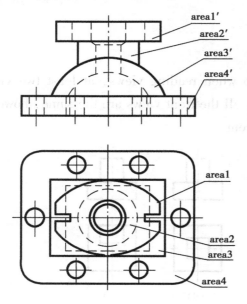

Fig. 5.22 Given views of a valve cover.

Example 5.4 The front and top views of a valve cover (阀盖) are shown in Fig. 5.22. Conceive the valve cover and supplement the left view of it.

Step 1 Divide areas in the front view and find out the corresponding views.

First, the front view is divided evidently into four areas enclosed by thick lines, which are areas 1′, 2′, 3′ and 4′ as shown in Fig. 5.22.

Based on the projection rules that the front and top views are aligned vertically, the corresponding view in the top view for each area can be found out one by one, which are area 1, area 2, area 3 and area 4.

Step 2 Visualize each primary object.

Based on area 1′ and area 1, object Ⅰ can be visualized as a top plate (顶板) with a hole in the middle and two slots located at left and right ends, respectively (Fig. 5.23(a)).

Based on area 2′ and area 2, object Ⅱ can be visualized as a vertical cylinder sleeve in which the upper is a conical hole while the lower is a cylindrical hole (Fig. 5.23(b)).

Based on area 3′ and area 3, object Ⅲ can be visualized as a partial cylinder with a through hole from the top to an empty cabinet (内腔) (Fig. 5.23 (c)).

Based on area 4′ and area 4, object Ⅳ can be visualized as a rectangular motherboard with four round corners, six small holes with the same diameter, and a rectangular hole in the middle (Fig. 5.23(d)).

Step 3 Assemble these primary objects together to form the valve cover.

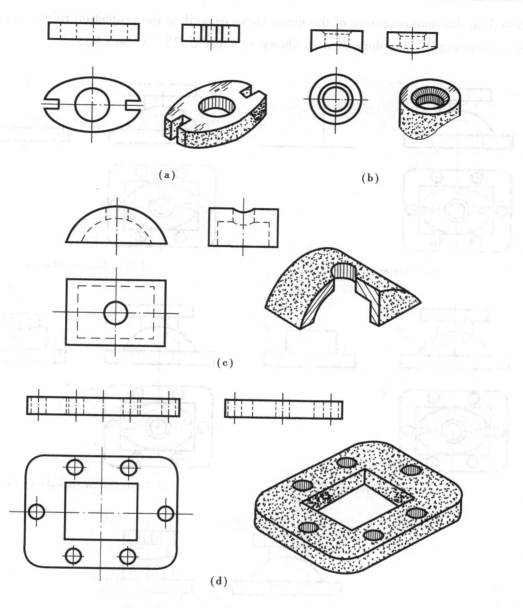

Fig. 5.23 Primary objects in the valve cover.

From the front and top views, it is known that above the motherboard, the partial cylinder is added, then the vertical cylinder sleeve is intersected, and lastly the top plate is added. Note that intersections exist between the partial cylinder and vertical cylinder sleeve. The two external cylindrical surfaces intersect to form outside intersection while the two internal cylindrical surfaces intersect to form inside intersection. Again, the internal cylindrical surface does not intersect the external cylindrical surface except at the entrance as they do not meet together in internal cabinet (内腔). Fig. 5.24 shows pictorial drawing of the valve cover.

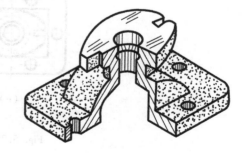

Fig. 5.24 Pictorial drawing of the valve cover.

Step 4 Supplement the left view.

Based on the understanding of the given views as well as the projection rules, the left view of the valve cover can be supplemented as shown in Figs. 5.25 (a) to (b).

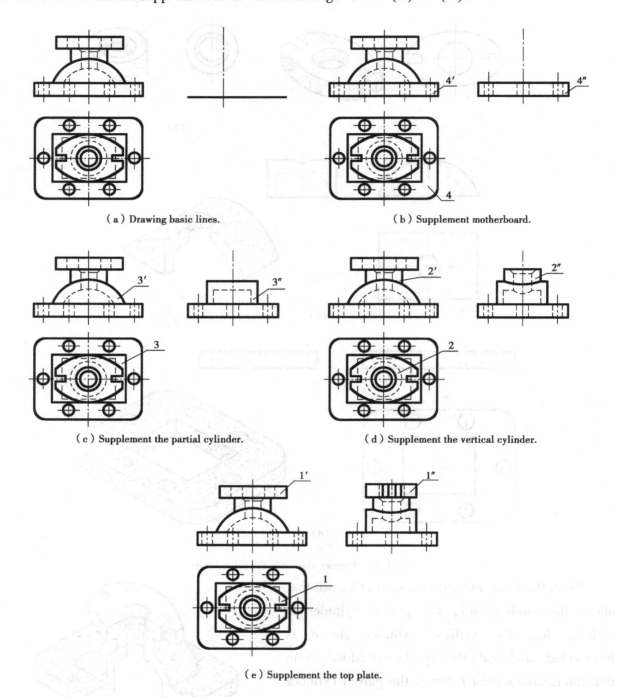

Fig. 5.25 Supplement left view.

1) Draw basic line, that is, the axis of the vertical cylinder sleeve and the base line of bottom surface (Figure (a)).

2) Supplement the left view of the motherboard—area 4″. Height of the area 4″ is aligned with area 4′ and the depth of it equals that of area 4 (Figure (b)).

3) Supplement the left view of the partial cylinder—area 3″ whose height is aligned with area 3′ and whose depth is measured from area 3 as shown in Figure (c).

4) Supplement the left view of the vertical cylinder sleeve—area 2″. Note that the outside intersection of 2″ and 3″ should be drawn with a thick line however, inside intersection of 2″ and 3″ should be drawn with a hidden line as shown in Figure (d).

5) Supplement the left view of the top plate—area 1″ whose height is aligned with area 1′ and whose depth is measured from area 1 as shown in Figure (e).

Lastly, check, clean and brighten all the thick lines to complete the left view of the object.

5.4.2.2 Analyzing-lines-and-planes method (线面分析法)

As mentioned in 5.2.1, a composite object is formed by two ways of superposition and cutting. Generally, analyzing-shape method could be adopted when reading the views of a superposed composite object while analyzing-lines-and-planes method could be used when reading the views of a cut-object.

A cut-object is a solid of which the other parts are taken away by cutting off, so some surfaces and lines are retained to enclose into the object.

Analyzing-lines-and-planes method is based on the fact that an area in a view represents a "surface" in space while a line in a view may indicate a "surface" or a "line" actually in space.

Usually, the front view is one of the given views because it tells shape and feature mostly. So, dividing areas and checking lines are used to read the front view with the help of the other views.

Steps of reading views with analyzing-lines-and-planes method are as follows.

Step 1　Read areas in the front view.

The front view is divided first into many areas including its external frame and then the corresponding projection of each area in the other view can be found out according to the projection rules. Finally, it is expected to figure out the spatial position of each surface shown by the given views.

Step 2　Read lines in the front view.

1) Pick out various lines in the front view.

Pick out lines parallel to axis OX, which may be horizontal planes or lines perpendicular to plane W in space.

Pick out lines parallel to axis OZ, which may be profile planes or lines perpendicular to plane H in space.

Pick out lines oblique to axis OX, which may be planes perpendicular to plane V or oblique lines in space.

2) Find out the corresponding projection of each line in the other view according to the projection rules.

3) Figure out the spatial meanings of each plance or each line shown by the given views.

Step 3　These surfaces and lines are assembled together to form a composite object.

Example 5.5　The front and left views are shown in Fig. 5.26 (a). Try to visualize the object and supplement the top view of it.

Step 1 Read areas in the front view.

Including the external frame (外框), the front view is divided into five areas which are areas 1', 2', 3', 4' and external frame 5' as shown in Fig. 5.26 (b).

Find out the corresponding projection of each area in the left view according to the projection rules, according to which the front and left views are aligned horizontally (高平齐). The corresponding left views of those areas are line segments 1", 2", 3", 4" and 5". Note that line segment 4" is in front of line segment 5" because area 4' is smaller than area 5'. Among these corresponding left views, only 2" is oblique line segment, and others are parallel to the axis OZ. So, it is observed that plane (2', 2") is perpendicular to plane W while planes (3', 3"), (1', 1"), (4', 4") and (5', 5") are parallel to plane V, frontal planes.

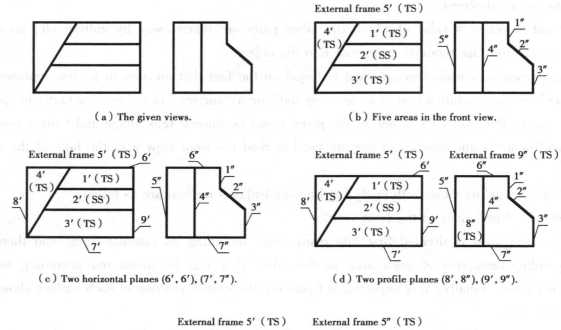

Fig. 5.26 The given views.

Step 2 Read lines in the front view.

1) Pick out lines parallel to axis OX. Although there are four lines parallel to axis OX in the front view, only two horizontal planes exist in the object because there are two corresponding line segments parallel to axis OY_W in the left view. As shown in Fig. 5.26 (c), lines 6' and 7' in the front view are located at the upmost and lowermost positions respectively. Corresponding left view of 6' is also line 6" parallel to axis OY_W. Similarly, corresponding left view of 7' is also line 7" parallel to axis OY_W. So, planes (6', 6") and (7', 7") satisfy projection characters of horizontal

planes and they are the top and bottom surfaces of the object.

2) Pick out lines parallel to axis OZ (Fig. 5.26 (d)). There are two lines, 8' and 9', parallel to axis OZ in the front view. According to the projection rules, their corresponding left views are area 8″ and external frame 9″. Evidently, planes (8', 8″) and (9', 9″) are profile planes which are left side surface and right side surface of the object. Notice that both area 8″ and the external frame of the left view, 9″, are true surfaces (TSs).

3) Pick out lines oblique to axis OX (Fig. 5.26 (e)). There is only one oblique line, 10', in the front view. According to the projection rules, its corresponding left view is area 10″. It is evident that plane (10', 10″) satisfies the projection character of the plane perpendicular to plane V. So, area 10″ is a foreshortened surface (FSs).

According to the previous analysis, the object is enclosed by ten surfaces, which are four frontal planes, two horizontal planes, two profile planes, one plane perpendicular to plane W and one plane perpendicular to plane V. Pictorial drawing of the object is shown in Fig. 5.27.

Step 3 Supplement the top view.

Based on the understanding of the given views and the object conceived, the top view can be supplemented as shown in Figs. 5.28 (a) to (d).

Before supplementing the top view, let's discuss how many areas are reflected in the top view. The horizontal plane appears in true size (TSs) in the top view while the plane perpendicular to plane V or plane W appears as foreshortened surfaces (FSs) in the top view. So, there are four areas in the top view, which are two TSs and two FSs.

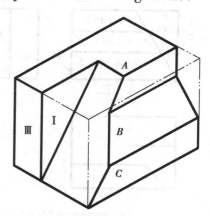

Fig. 5.27 Pictorial of the composite object.

The processes of supplementing the top view are as follows.

1) Draw the top view of the complete rectangular prism, a rectangular frame whose width is aligned with the front view and the depth is measured from the left view (Fig. 5.28 (a)).

2) Supplement four convergent lines of frontal planes which are line segments 3, 1, 4 and 5 as shown in Fig. 5.28 (b). Among them, line segments 3 and 5 have existed at the front side and the rear side of the rectangular frame. Take line segment 1 as an example of how to obtain it. The length of line segment 1 in left-right direction is aligned with area 3'. The distance from line segment 3 to the rear surface is same with that from line segment 3″ to the rear surface. Similarly, line segment 4 can be obtained.

3) Supplement foreshortened surface of the plane (2', 2″). Plane (2', 2″) is perpendicular to plane W and it contains two lines perpendicular to plane W, which are lines ($e'f'$, $e''f''$) and ($g'h'$, $g''h''$). Based on e' and e'', the top view point e can be obtained. Similarly, the top views f, g, h can be got. Plane (2', 2″) appears as FS area e'-f'-h'-g'-e' in the front view and as FS area e-f-h-g-e in the top views (Fig. 5.28 (c)).

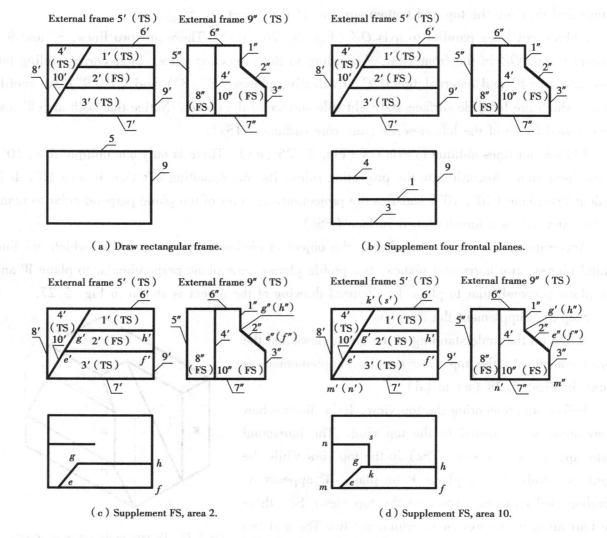

Fig. 5.28 Supplement the top view.

4) Supplement foreshortened surface of the plane (10'-10"). Plane (10'-10") is perpendicular to plane V and it contains two lines perpendicular to plane V, which are lines ($k's'$, $k''s''$) and ($m'n'$, $m''n''$). From k' and k'', the top view point k may be supplemented. Similarly, the top views, s, n, m, may be obtained. Plane (10', 10") appears as FS area $k''\text{-}s''\text{-}n''\text{-}m''\text{-}e''\text{-}g''\text{-}k''$ in the left view and as FS area $k\text{-}s\text{-}n\text{-}m\text{-}e\text{-}g\text{-}k$ in the top view (Fig. 5.28 (d)).

In the top view, the true size of the top surface of the object is area 6 while the outer frame of the top view is the true size of the bottom of the object.

Finally, check, erase unnecessary lines and brighten the visible lines to obtain the top view of the object.

Chapter 6　Views

Sometimes the three views are not enough to represent the shape and structure clearly of a very complicated object. It is necessary to use some other representing methods in order to make drawings more clear and easy to undenstand. According to the Chinese National Standard of Technical Drawing (GB/T 17451—1998, GB/T 17452—1998), the representing methods are as follows: principal views (基本视图), removed views (向视图), partial views (局部视图) and auxiliary views (斜视图) as well as sectional views (剖视图) and so on. This chapter will deal with views used to show the external features of an object.

6.1　Principal Views (基本视图)

6.1.1　Principal projection planes

The Chinese National Standard of Technical Drawing stipulates that the principal projection planes used in engineering drawings are six planes of a cube box (Fig. 6.1).

6.1.2　Principal views

As discussed above, suppose that an object is put properly in six projection planes and the object is projected toward them to generate six principal views-front view, top view, left view, rear view (后视图), bottom view (仰视图) and right view (右视图) (Fig. 6.2). Six surfaces on the object are shown clearly by six views. Namely, the frontal surface of the object is represented by front view; the top surface by top view; the left side by left view, the bottom surface by bottom view, the rear surface by rear view and the right side by right view.

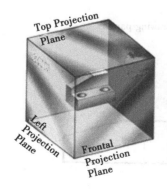
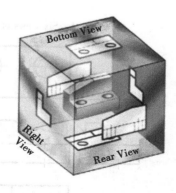

Fig.6.1　Principal projection planes.　　　Fig.6.2　Principal views.

When the six principal views are obtained, the object is removed out of the six projection planes, but the six views are yet located in the six principal projection planes perpendicular to

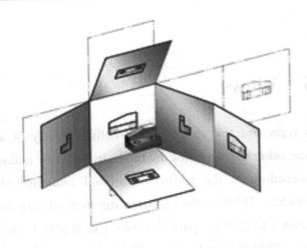

Fig. 6.3 Unfolding of the box.

each other. To show six principal views on a flat paper, it is necessary to unfold the six projection planes. Assuming that four of the six projection planes, surrounding plane V, are hinged to the plane V, the cube box is unfolded starting from the left edge of it (Fig. 6.3) and each projection plane revolves outwardly from its original position until they lie in the plane V as shown in Fig. 6.4. Arrangement of the six principal views is stipulated by the Chinese National Standard of Technical Drawing. The bottom, front and top views are aligned vertically while the right, front, left and rear views are aligned horizontally. To draw a view out of the arrangement is a serious error in drawing. In the case as shown in Fig. 6.4, the names of six principal views are omitted (Fig. 6.5). In addition, the arrangement according to the Chinese National Standard is also called standard arrangement.

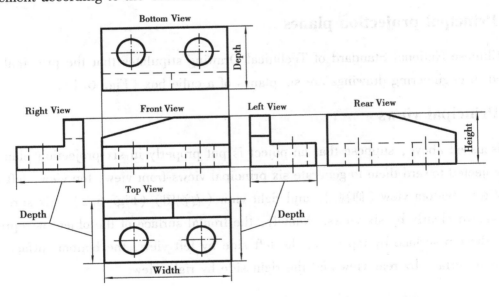

Fig. 6.4 Six views of an object on a sheet of drawing paper.

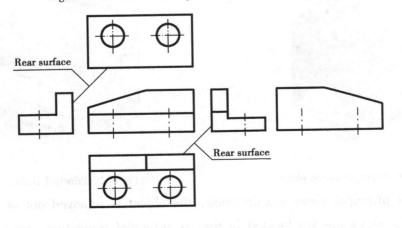

Fig. 6.5 Six Principal Views.

6.1.3 Projection rules of an object

Based on the generation process of the six views, the object's projection rules are obtained, which are used to designate the position relationship among the six principal views.

The bottom, front, and top views are aligned vertically to show the longitude of the object (长对正). Besides, rear view also represents the width of the object.

The right, front, left, and rear views are aligned horizontally to show the latitude of the object (高平齐).

The top, left, bottom, and right views have the same depth of the object (宽相等).

Note that in these four views that surround the front view, i.e. top view, bottom view, left view as well as right view, the rear surfaces of the object face towards the front view (Fig. 6.5).

Usually, views are used to express the external features of an object. So, hidden lines in the views may be omitted except otherwise specified (Fig. 6.5).

6.1.4 Projection systems

6.1.4.1 First-angle projection

As stated in 2.2, first-angle projection is used in our country. It shows the right view to the left of the front view and the top view under the front view. The truncated cone is the symbole that designates first-angle projection (Fig. 6.6).

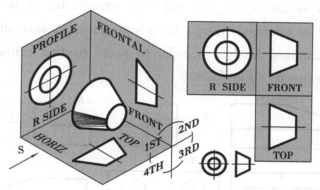

Fig. 6.6 Symbol used to designate first-angle projection.

Third-angle projection is used for drawing orthographic views in the U.S., Great Britain and Canada. The top view is placed over the front view and the right view is placed to the right of the front view. The truncated cone is the symbole that designates third-angle projection (Fig. 6.7). As explained in 2.2, in third-angle projection, projection planes are assumed to be transparent so that the observer can see the object through the plane.

When metric units of measurement are used, the SI symbol is given in combination with the truncated cone on the drawing.

Symbol: **SI▷⊕** is placed on drawings to specify first-angle projection and metric units of measurement.

Fundamentals of Engineering Drawing

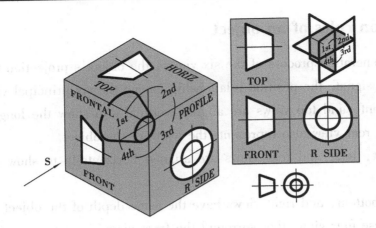

Fig. 6.7 Symbol used to designate third-angle projection.

Symbol: ⬚ is placed on drawings to specify third-angle projection and metric units of measurement.

6.2 Removed Views (向视图)

6.2.1 Definition

A removed view is similar to a principal view except that it is not placed in standard arrangement but instead is arranged freely. Its advantage is to save drawing space because a removed view may be arranged freely. In order to make reading convenient and not to make confusion, labeling is necessary (Fig. 6.8).

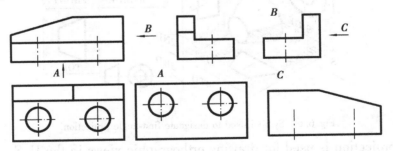

Fig. 6.8 Removed views.

6.2.2 Label

A label consists of arrow, capital letter, and the name of the removed view.

Arrow indicates the direction of the sight.

Any **capital letter** such as A or B is used to name the arrow. The named arrow is called arrow A or arrow B.

Name of the removed view is specified by the corresponding capital letter A or B. The removed view is called view A or view B.

As shown in Fig. 6.8, view A is actually the bottom view, view B is the right view and view C is the rear view.

Note ①. In general, the front, top and left views should not be arranged freely while the right, bottom and rear views may be drawn by removed view respectively.

Note ②. Usually the arrow indicating the direction of sight should be placed surrounding the front view as arrows A and B shown in Fig. 6.8, while the arrow indicating the direction from front to rear can only be put near the left view or right view and the arrow should be pointed to rear side of the object as arrow C shown in Fig. 6.8.

6.3 Partial Views (局部视图)

6.3.1 Definition and application

A partial view is a view produced by projecting a partial object towards to any principal projection plane. It is applied when only part of object is necessary to be described. For example, in Fig. 6.9, the main part of the object consists of three vertical coaxial (同轴的) cylinders which can be represented clearly by the front and top views, except the left and right sides of the object. However, to avoid expressing the coaxial cylinders repeatedly, partial views are used. When partial left end structure of the object is projected to the right-profile projection plane, the partial left view is obtained. Similarly, when partial right end structure of the object is projected to the left-profile projection plane, the partial right view, view B, is obtained.

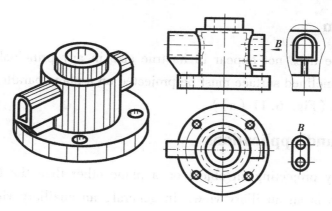

Fig. 6.9 Partial view of an object.

6.3.2 Wave line in a partial view

A break line, wave line, is used to limit the partial view. It is a projection of a broken surface of the object and drawn with HB pencil. When the contour of certain surface is closed by itself, the break surface does not exist and do not draw it by wave line. View B in Fig. 6.7 is an example.

6.3.3 Label of a partial view

Label of a partial view is the same as that of a removed view, but, if a partial view is arranged in the standard arrangement of views, the label may be omitted. For example, in Fig. 6.9, partial left view needn't to be labeled.

Note that a partial view of a symmetric object may be drawn as follows. First, only half view or a quarter of view needs to be drawn at one side of the centerline. Secondly, two short thin parallel lines perpendicular to symmetric centerline are drawn at the two ends of symmetric centerline (Fig. 6.10). Fig. 6.8 exhibits a partial view, in which the wave line of broken boundary are substituted by symmetric centerlines.

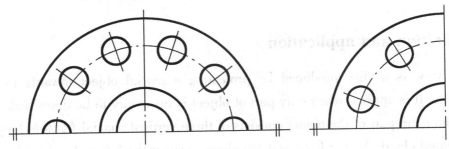

Fig. 6.10 A special partial view.

6.4 Auxiliary Views (辅助视图)

6.4.1 Introduction

An inclined surface does not appear in its true size in any principal projection plane. To show the true size, the inclined surface must be projected on a plane parallel to it. As a result, an auxiliary view is formed (Fig. 6.11 (a)).

6.4.2 Definition and applications

A view obtained by projecting an object on a plane other than the horizontal, frontal and profile projection planes is an auxiliary view. In general, an auxiliary view is projected onto a plane that is perpendicular to one of the principal projection planes but parallel to the inclined structure of the object.

The object shown in Fig. 6.11 (a) has an inclined surface P that does not appear in its true size in any principal view. To solve the problem, an auxiliary projection plane H_1 is introduced which is parallel to the surface P but perpendicular to the plane V and hinged to it. When projecting along arrow A, the auxiliary view A is obtained. Thus when plane H and auxiliary plane H_1 are unfolded to lay them on a drawing paper. The folding lines are defined as projection axis OX and axis OX_1 (Fig. 6.11 (b)). The drawing is simplified by retaining the projection axes and omitting the projection planes (Fig. 6.11 (c)). As shown in Fig. 6.9 (c), distances Y_1 in both

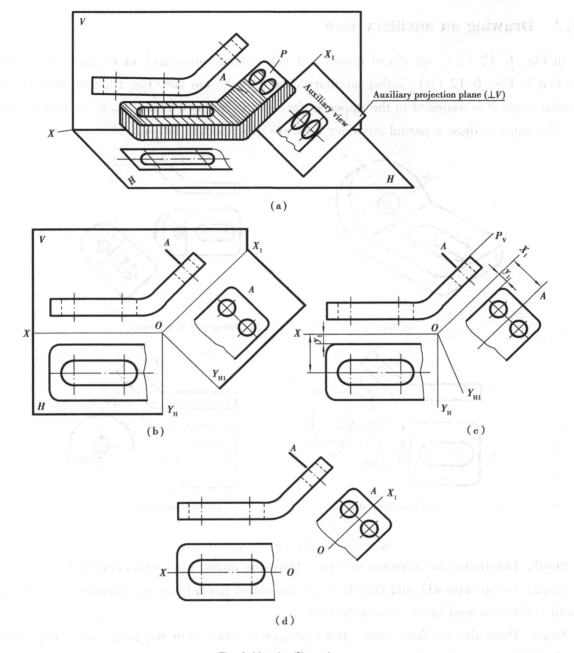

Fig. 6.11 Auxiliary views.

partial top view and auxiliary view must be equal since both of them represent the distance between the rear surface of the object and plane V. In this case, axes OX and OX_1 are regarded as the measurement datum. Besides, axes OX and OX_1 may be moved forward. For example, they may coincide with the rear surface of the object or may pass through the symmetrical plane as shown in Fig. 6.11 (d). Lastly, it should be remembered that axis OX_1 is always parallel to inclined surface. In Fig. 6.11 (c), P_V is frontal piercing line (正面迹线) of the inclined surface P, thus axis OX_1, is parallel to P_V.

In general, auxiliary view is a partial view. When an inclined surface is represented clearly, the inclined surface is broken with a wave line. View A in Fig. 6.11 is an example.

6.4.3 Drawing an auxiliary view

In Fig. 6.12 (a), an object consists of a horizontal plate and an inclined plate. If it is described in Fig. 6.12 (b), either auxiliary view A or the top view has a foreshortened surface. If capital letter P is assigned to the upper surface of the inclined plate, P_V is its frontal piercing line. The steps to draw a partial auxiliary view are described as follows.

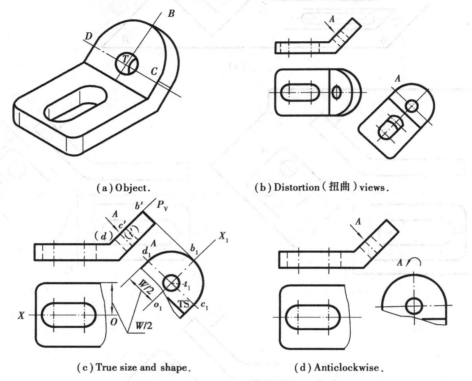

Fig. 6.12 Drawing a partial auxiliary view.

Step1. Determine the direction of sight. Draw the arrow A perpendicular to P_V.

Step2. Locate axes OX and OX_1. Because the object has a front-rear symmetrical plane, axes OX and OX_1 are located in the symmetrical plane.

Step3. Draw the auxiliary view. Draw projection lines from the front view. For example, through b' draw projection line to intersect axis OX_1 at b_1 and from $c't_1$ draw projection line to intersect axis OX_1 at t_1. Transfer dimension W/2 from the top view to auxiliary view with divider to intersect the projection line from c' (d') at c_1 and d_1, respectively. Note that draw the projection line from c' (d') in center line since it is the symmetrical centerline of the revolutionary surface. With point t_1 as center and distance $t_1 d_1$ as radius, draw semicircle. From c_1 and d_1 draw lines parallel to axis OX_1 and the auxiliary view A is limited by a break line, i.e. a wave line. Besides, a small hole is drawn at point t_1 and the radius is measured from the front view.

Finally, check, clean-up the unnecessary lines and brighten all the visible lines with proper pencil of B or 2B.

Fig. 6.13 shows another example with respect to the auxiliary view. The steps for drawing an auxiliary view are the same with those Fig. 6.12. Evidently, axis OX_1 may yet be located at the front-rear symmetrical plane. In order to describe the object clearly and simply, the front view, partial auxiliary view A, and partial left view as well as partial top view are used (Fig. 6.14).

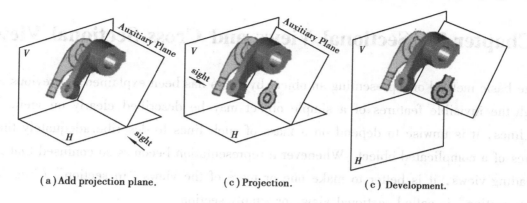

Fig. 6.13 Steps of drawing an normal view.

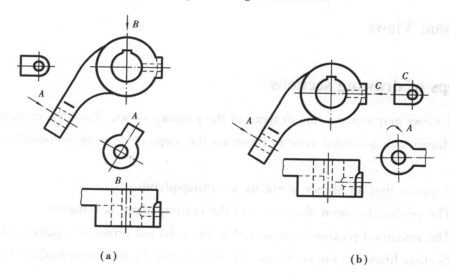

Fig. 6.14 A partial view and an auxiliary view.

6.4.4 Labeling an auxiliary view

An auxiliary view must be labeled. Labeling an auxiliary view is as same as a removed view, but it should be pointed out that either the capital letter next to the arrow or the capital letter above the auxiliary view should be written uprightly (正写)—that is, the head of the capital letter should be upward. Note that the arrow must be perpendicular and pointed to the inclined part of the object (Fig. 6.14). Usually, an auxiliary view is laid in the direction of sight. But it may be out of the direction of sight and rotated into normal position (常规位置). In this case, a rotating symbol must be added. Drawing a rotating symbol is shown in Fig. 6.15.

Note that the arrow head must face the corresponding capital letter. In Figs. 6.12 (d) or 6.14 (b), view A is the example of how to use a rotating symbol.

Fig. 6.15 Rotating symbol.

Chapter 7 Sectional Views and Cross-Sectional Views

The basic method of representing an object by views has been explained in previous chapters. Although the invisible features of a simple object may be described clearly on views by using hidden lines, it is unwise to depend on a mass of such lines to describe adequately the interior structures of a complicated object. Whenever a representation becomes so confused that it prevent from reading views, it is better to make one or more of the views "in section" (Fig. 7.1). A view "in section" is called sectional view, or simply section.

7.1 Sectional Views

7.1.1 Steps of drawing sections

Sectional views may replace one or more of the primary views. Take the example of which the front view is drawn as a sectional view to illustrate the steps of drawing sectional view (Figs. 7.1 (a) to (c)).

Step1 Suppose that an object is cut by a cutting-plane.
Step2 The portion between observer and the cutting-plane is removed.
Step3 The remained portion is projected to the selected projection plane, plane V.
Step4 Section lines are drawn across the surface cut by the cutting-plane to emphasize the solid part of the interior.
Step5 Labeling.

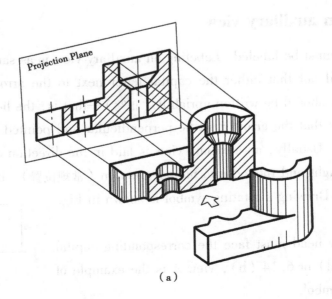

(a)

Chapter 7 Sectional Views and Cross-Sectional Views

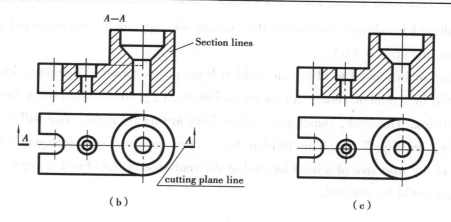

Fig. 7.1 A section.

Concepts with regard to sectional view are explained as follows.

7.1.1.1 Cutting-plane

In general, cutting-plane passes through symmetric plane of an object and parallel one of the projection planes. For example, in Fig. 7.1, the cutting-plane is parallel to plane V while in Fig. 7.2, the cutting-plane A and cutting-plane B are parallel to plane W, respectively.

The cutting-plane is indicated in the view adjacent to the sectional view. In this view, the cutting-plane appears as a line called cutting-plane line (剖切平面线) which is represented by two short thick line segments about 3-5mm or a center line as shown in Fig. 7.1(b) and Fig. 7.2.

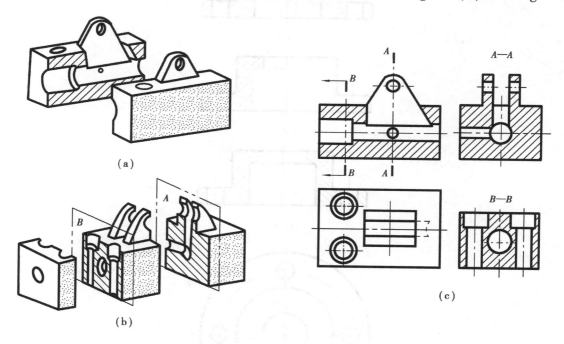

Fig. 7.2 Cutting-planes.

7.1.1.2 Projecting

In Fig. 7.3, an object is cut by a cutting-plane parallel to plane V, of which the correct front sectional view is shown in Fig. 7.2 (a).

When the remained portion is projected to plane V, all visible edges and contours behind the cutting-plane should be shown, otherwise the section will appear as disconnected and unrelated parts, as shown in Fig. 7.3 (b).

    Sections are mainly used to eliminate hidden lines in the views. Therefore, hidden lines are generally omitted in sectional views. As shown in Fig. 7.3(c), the hidden lines do not clarify but confuse the drawing. However, sometimes hidden lines are useful if their use will make it possible to omit a view. In Fig. 7.4, The hidden lines used in the front sectional view make the representation of the true size of a boss located at the rear side of the object is represented clearly; thus a rear view could be omitted.

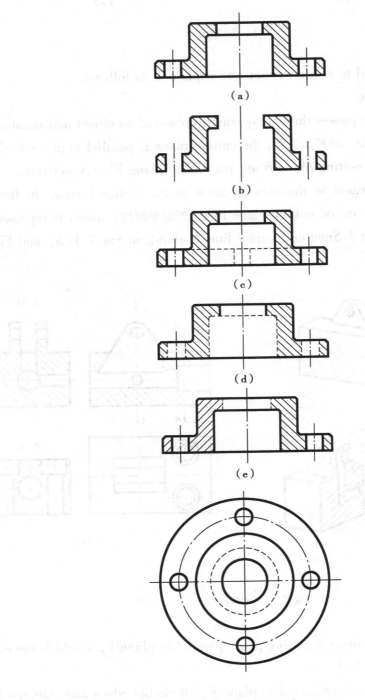

Fig. 7.3 Lines in sections.

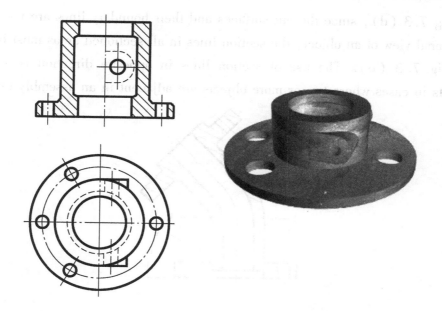

Fig. 7.4 Hidden lines in section.

7.1.1.3 Section line (剖面线)

Section line symbols as shown in Fig. 7.5 are used to indicate specific material, such as cast iron, steel, brass and so on. However, the section line symbol of castiron may be used to represent any material, so it is generally used for all materials and called section line simply.

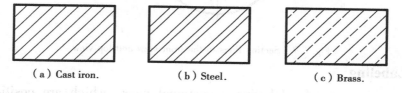

(a) Cast iron.　　　(b) Steel.　　　(c) Brass.

Fig. 7.5 Symbols for section line.

Draw section line with thin line at an angle of 45° with horizontal line as shown in Fig. 7.6 (a). Space between the section lines is as evenly as possible by eye from approximately 1.5 mm to 3 mm or more apart, depending on the sectioned area (close together in small areas and farther apart in larger areas).

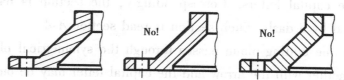

Fig. 7.6 Direction of section lines.

If section lines in a section view at 45° with horizontal would be parallel or perpendicular (or nearly so) to a prominent (重要的) visible outline, the angle can be changed to 30° or 60°, however, in other section views, section lines are drawn at angle of 45° with horizontal line (Fig. 7.6). For example, as shown in Fig. 7.7, section lines in the front view are drawn at 30° with horizontal, but in the top sectional view, section lines are drawn at 45° with a horizontal line.

Besides, a section-lined area is always completely bounded by thick lines, never hidden lines, as in Fig. 7.3 (d), since the cut surfaces and their boundary lines are visible.

In a sectional view of an object, the section lines in all sectioned areas must be parallel, not as shown in Fig. 7.3 (e). The use of section lines in opposite direction is an indication of different objects in cases where two or more objects are adjacent in an assembly drawing.

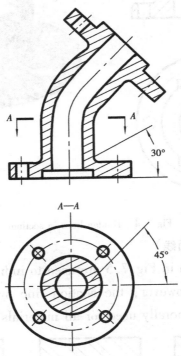

Fig. 7.7 Section lines in the front view and top view.

7.1.1.4 Labeling

Four items are necessary for labeling a sectional view, which are position of the cutting-plane, projection direction, the name of the cutting-plane and the name of the sectional view.

As shown in Fig. 7.1(b), the position of the cutting-plane is indicated by cutting-plane lines and they are drawn at the adjacent view of the section. The projection direction of the section is represented by arrowheads, which are perpendicular to the cutting-plane lines. The cutting-plane is named by two same capital letters. Correspondingly, the section is named by the two same capital letters separated by a dash. Such section is read section *A-A*.

However, if only one cutting-plane passes through the symmetrical plane of the object, the cutting-plane line, together with the arrow and the capital letter may be omitted (Fig. 7.1(c)).

7.1.2 Types of sectional views

There are three types of sectional views: full section (全剖视图), half section (半剖视图) and broken-out section (局部剖视图).

7.1.2.1 Full section

A full section is one in which the cutting-plane passes fully through an object, so that the resulting view is completely "in section". A full section shows the object's internal features clearly. The sections mentioned previously are full sections as shown in Figs. 7.1, 7.2, 7.3

(a),7.4,7.7.

7.1.2.2 Half section

A half section is generally used to represent symmetrical objects in which one half is drawn in section and the other half as an exterior view to show both external and internal features. Note particularly that a center line separates the exterior and interior portions in the half section view. It is necessary to omit hidden lines in a half section. Labeling a half section is identical to full section.

Fig. 7.8 (a) shows the front and top views of an object with all its hidden lines. The object can be better described by half sections in the front view and top view respectively. From the given views, it is known that the object consists of upper plate, vertical cylinder sleeve and lower plate.

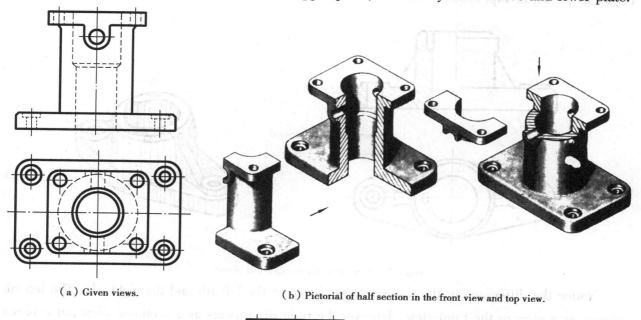

(a) Given views.　　　　(b) Pictorial of half section in the front view and top view.

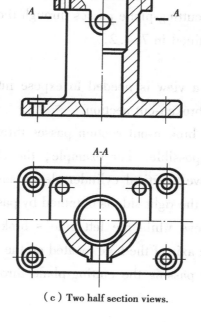

(c) Two half section views.

Fig. 7.8　Half sections.

As shown in Fig. 7.8 (b), suppose that the cutting-plane passes halfway through the symmetrical plane of the object and removes a quarter of the object, the resulting view, front view, is a half section in which the left half is an exterior view and the right half is section (Fig. 7.8(c)). As shown in Fig. 7.8 (c), suppose that the cutting-plane A-A passes through the axis of the two small holes located on the front half and rear half of the vertical cylinder sleeve. At the same time, the front half above the cutting-plane A-A is removed. The resulting view, the top view, is a half section in which the rear half is an exterior view and the front half is a section (Fig. 7.8 (d)).

Besides, if an object is approximately symmetrical, half section is also applicable to this object as shown in Fig. 7.9.

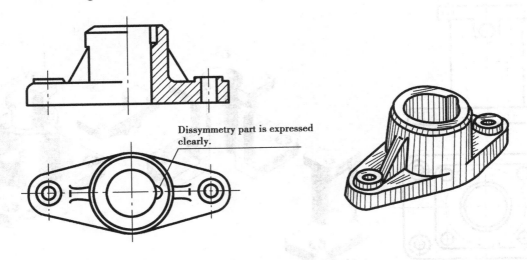

Fig. 7.9 Approximately symmetrical object.

Notice the difference in the representation between the left rib and the right rib. The left rib appears as a view in the front view, however the right rib appears as a sectional view but it is not sectioned in the front view since the cutting-plane passes through the rib longitudinally (纵向地). The representation of the rib is explained in 7.4.2.

7.1.2.3 Broken-out section

When only a partial section of a view is needed to expose interior shape, such a section, limited by a break line, is called a broken-out section.

Usually, the cutting-plane in a broken-out section passes through the hole's axis to expose the interior structures as much as possible. For example, the object shown in Fig. 7.10 is composed of horizontal cylinder sleeve, vertical cylinder sleeve and a motherboard. In the front view, the broken-out section located the right side is obtained by passing the cutting-plane through the axis of the vertical cylinder sleeve while the left side's broken-out section is obtained by passing the cutting-plane through the axis of the hole located on the motherboard. In the top view, the broken-out section is formed by passing the cutting-plane through the axis of the horizontal cylinder sleeve.

Chapter 7 Sectional Views and Cross-Sectional Views

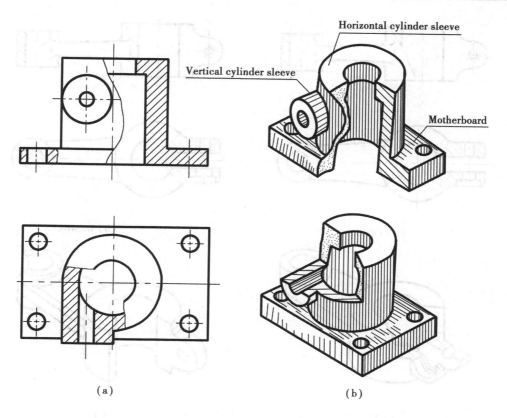

Fig. 7.10 Broken-out section.

A broken-out section is generally not labeled. That is to say, if broken-out section is obtained by passing a cutting-plane through the axis of the hole, the corresponding cutting-plane line is omitted.

Besides, broken-out sections may be used in the unsectioned view in a half section. For example, in Fig. 7.8 (d), in order to represent the through holes in the upper plate and lower plate respectively, two broken-out sections are applied in the view unsectioned in the front view.

Pay attention to the following points about broken-out section.

1) In a hole, there is no wave line as it is empty. What's more, the wave line should not be beyond contours of the object. As shown in Fig. 7.11, figure (a) is correct while figure (b) is incorrect. Pictorial drawing of the broken-out section in the top view is shown in figure (c). Pictorial drawing of the broken-out section in the front view is shown in figure (d).

2) If a cylindrical hole is represented by broken-out section, the hole's axis can be used as boundary line of section (Fig. 7.12). However, a prismy hole is represented by broken-out section, only wave line can be used as boundary line of the section (Fig. 7.13). If thick lines exist in the symmetrical plane of the object, half section is not applicable. So, a broken-out section is an alternative as shown in Fig. 7.14.

143

Fundamentals of Engineering Drawing

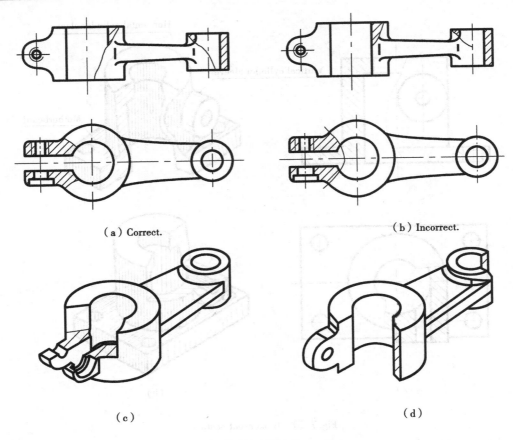

Fig. 7.11 Wave lines.

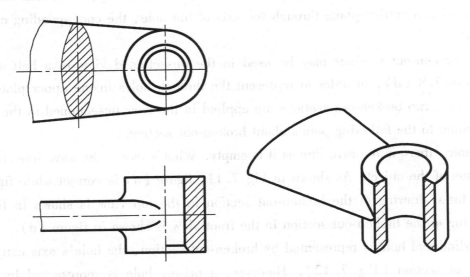

Fig. 7.12 Broken-out section in a revolution.

3) For convenience in computer drawing, long break line is used as shown in Fig. 7.15. Long break lines can be extended beyond the contours of the object.

4) The wave line in broken-out section should not be omitted, so Fig. 7.16 (b) is incorrect. The wave line should not meet with other lines at their endpoints as shown in Fig. 7.16 (c). In other words, the wave line is not the extension line of any line.

Chapter 7 Sectional Views and Cross-Sectional Views

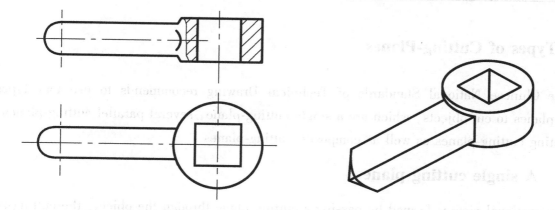

Fig. 7.13 Broken-out section in a rectangular prism.

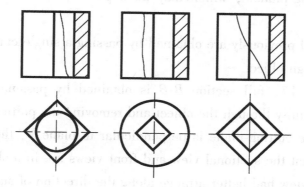

Fig. 7.14 Broken-out sections in symmetrical objects.

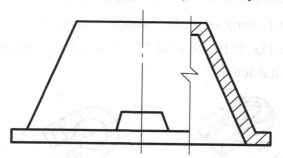

Fig. 7.15 Long break line.

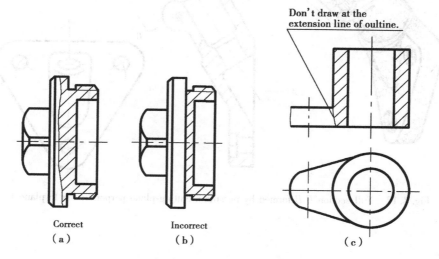

Correct Incorrect
(a) (b) (c)

Fig. 7.16 Break lines in broken-out sections.

145

7.2 Types of Cutting-Planes

The Chinese National Standards of Technical Drawing recommends to use four types of cutting-planes to cut objects, which are a single cutting-plane, several parallel cutting-planes and intersecting cutting-planes as well as composite cutting-planes.

7.2.1 A single cutting-plane

If a sectional view is formed by passing a cutting-plane through the object, the cutting-plane is called "a single cutting-plane", which may be a plane parallel or perpendicular to one of projection planes.

Those sections stated previously are obtained by passing a single cutting-plane parallel to one of projection planes through object.

As shown in Fig. 7.17, full section B-B is obtained by passing the cutting-plane B, B perpendicular to plane V fully through the object and removing the portion below the cutting-plane B, B. Note that when the cutting-plane is perpendicular to plane V, the resulting sectional view must be rotated 90° so that the sectional view and front views are in a sheet of paper. Under this condition, the sectional view had better arrange along the direction of sight. However, in order to draw conveniently, sometimes the sectional view may be removed out of the direction of sight and rotated into the position of a primary view, for example section B-B ⌒ as shown in Fig. 7.18 and section A-A ⌒ as shown in Fig. 7.19. Symbol "⌒" is called rotating symbol in which the arrow head must be faced the capital letter.

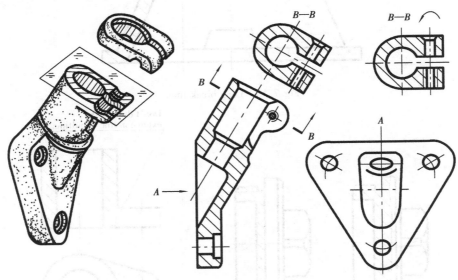

Fig. 7.17 Full section B-B formed by passing a cutting-plane perpendicular to plane V.

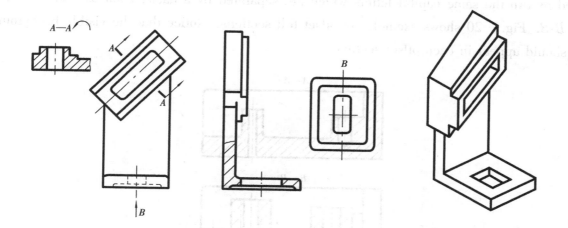

Fig. 7.18 Broken-out section A-A formed by a cutting-plane perpendicular to plane V.

7.2.2 Several parallel cutting-planes

A sectional view is formed by passing several parallel cutting-planes through object in which several interior features are not in a single plane. Such a section is called an offset section in the U. S engineering drawing books. Note that these parallel cutting-planes are parallel to one of the projection planes, such as parallel to plane V, plane H or plane W.

In Fig. 7.19 (a), offset full section A-A is formed by passing two parallel cutting-planes through the two interior features of the object. The two parallel cutting-planes are parallel to plane V and pass through the large hole and one of small holes of the object, respectively. It should be pointed out that turned plane between the two parallel cutting-planes is not shown in the section because the cutting is imaginary.

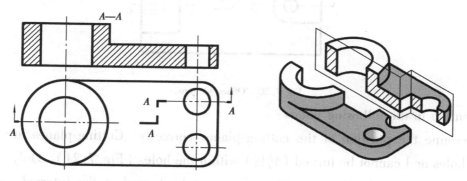

Fig. 7.19 Offset section.

Labeling for offset section is indispensable. Cutting-plane lines are drawn in the adjacent view of the offset section to represent the positions of cutting-planes. As shown in Fig. 7.19 (b), at places where the cutting starts, turns and ends, draw cutting-plane lines and write the same capital letters such as A or B. Two arrows indicating the direction of projection are drawn in the outside ends of the cutting-plane lines located outside the contours of the view. However, if the direction of projection is obvious, the arrows may be omitted. Correspondingly, the offset section is named by two the same capital letters which are separated by a dash, such as section *A-A* or section *B-B*. Fig. 7.20 shows examples of offset full sections. Notice that the visible background shapes should appear in each offset section.

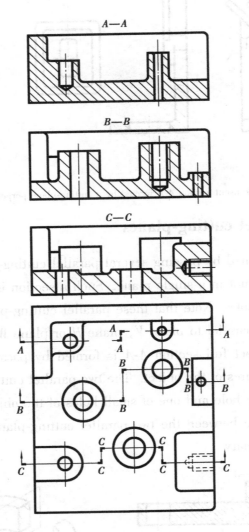

Fig. 7.20 Offset sections.

Pay attention to the following points:

(1) Determine the position of the cutting-planes correctly. Cutting-planes are supposed to pass through holes and cannot be turned (转折) within the holes (Fig. 7.21). Only when internal structures share a symmetric plane, cutting-plane can be turned at the internal structures. As shown in Fig. 7.22, the separating line between two internal structures is centerline.

(2) It is emphasized again that the turning of cutting-planes is theoretical and imaginary, so two edges, which would be implied by a thick line, do not exist in the section as shown in Fig. 7.23.

Chapter 7 Sectional Views and Cross-Sectional Views

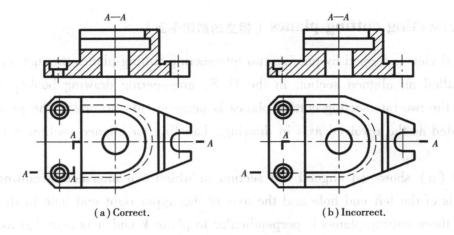

(a) Correct.　　　　　　　　(b) Incorrect.

Fig. 7.21　Selecting cutting-planes correctly.

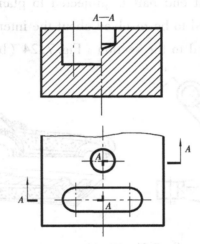

Fig. 7.22　Offset section for symmetrical structure.

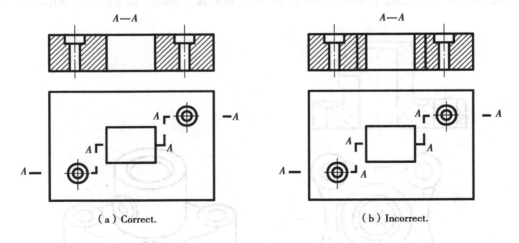

(a) Correct.　　　　　　　　(b) Incorrect.

Fig. 7.23　Offset section.

7.2.3 Intersecting cutting planes (相交的剖切平面)

A sectional view is formed by passing two intersecting cutting-planes through an object. Such a section is called an aligned section in the U.S. engineering drawing books. Note that the intersection of the two intersecting cutting-planes is perpendicular to one of the projection planes and it is regarded as the revolved axis in drawing. Labeling for aligned section is the same with offset section.

Fig. 7.24 (a) shows an aligned full section in which two intersection cutting-planes pass through the axis of the left end hole and the axis of the upper right end hole in the object. The intersection of those cutting-planes is perpendicular to plane V and it is regarded as the revolved axis. When the object is cut by the two intersecting cutting-planes and the portion above the two cutting-planes is removed, the left end part is projected to plane H to form the section directly while the right end part is imagined to be revolved about the intersection to the position parallel to plane H, from where it is projected to the section (Fig. 7.24 (b)).

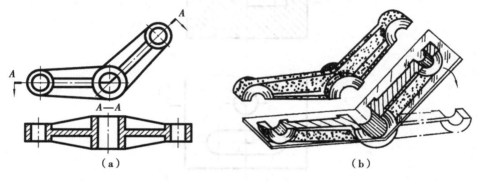

Fig. 7.24 Aligned section Ⅰ.

Fig. 7.25 shows another example of an aligned full section. From the top view, it is known that in two intersecting cutting-planes, the right side one is parallel to plane V while the left side

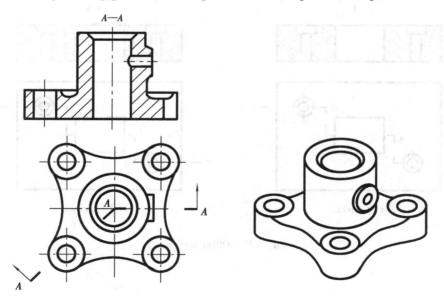

Fig. 7.25 Aligned section Ⅱ.

one is perpendicular to plane H and inclined to plane V. The intersection of those cutting-planes is perpendicular to plane H and it is regarded as the revolved axis. When the object is cut by the two intersecting cutting-planes and the portion in front of the two cutting-planes is removed, the object's right part is projected to plane V to form the section directly while the object's left part is imagined to be revolved about the intersection to the position parallel to plane V, from where it is projected to the section.

Aligned full section A-A as shown in Fig. 7.26 is analyzed by readers.

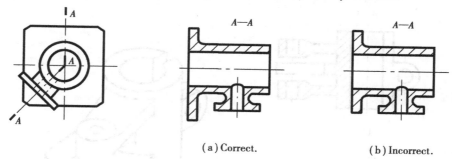

(a) Correct. (b) Incorrect.

Fig. 7.26 Aligned section Ⅲ.

Pay attention to the following points:

1) When cutting-plane is imagined to be revolved to the position parallel to a certain projection plane, only those structures that contact with the cutting-plane or have the close relation with the cutting-plane are revolved. However other structures are projected from their origin position. For example, in Fig. 7.27, boss (凸台) is not revolved and it is projected to the section directly.

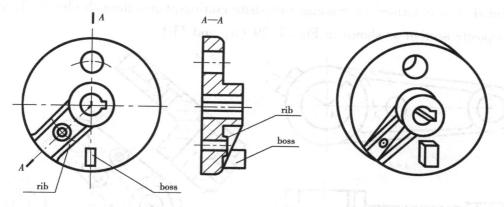

Fig. 7.27 Section cutting by intersecting cutting planes.

2) In Fig. 7.28, there are three arms in the right side of the object. The middle arm is a solid one (实心) while others have holes. Aligned full section A-A as shown in Fig. 7.28 (b) is real projection, in which the middle arm is cut partly. However the real projection may lead to misunderstand to solid arm. So, the middle arm is regarded as not to be sectioned as shown in Fig. 7.28 (a). Fig. 7.28 (d) is the pictorial drawing of the object.

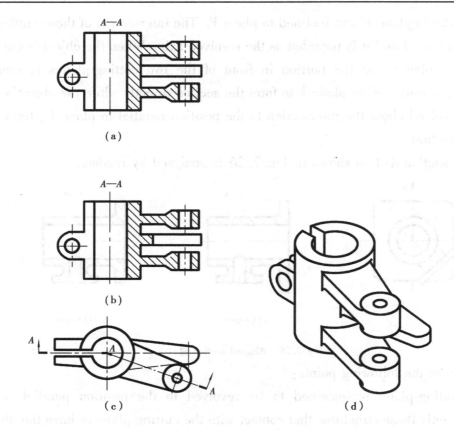

Fig. 7.28 Section by intersecting cutting planes.

7.2.4 Composite cutting planes（组合的剖切平面）

A sectional view is formed by passing composite cutting-planes through object. This section is called composite section as shown in Figs. 7.29 (a) and (b).

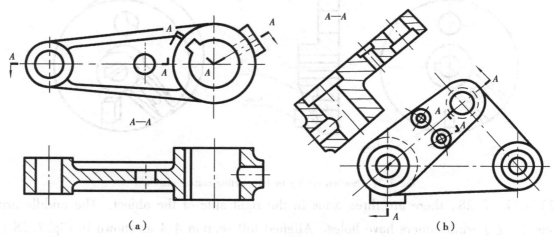

Fig. 7.29 Section by intersecting cutting planes.

In summary, no matter what kind of cutting-planes, a single cutting-plane, parallel cuttling-planes or intersecting cutting-planes are used, any sectional view-full section, half section or broken-out section may be obtained. For example, Fig. 7.30 shows a half section formed by intersecting cutting-planes. Fig. 7.31 shows a broken-out section formed by parallel cutting-planes.

Chapter 7 Sectional Views and Cross-Sectional Views

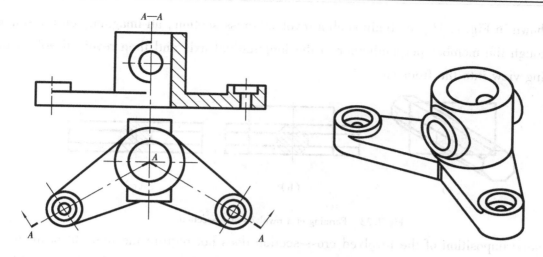

Fig. 7.30 A Half section formed by intersecting cutting planes.

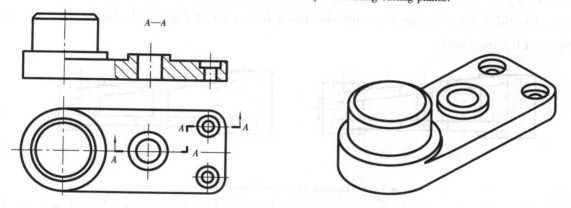

Fig. 7.31 A broken-out section by parallel cutting-planes.

7.3 Cross-Sections (断面图)

7.3.1 Revolved cross-sections (重合断面图)

A revolved cross-section is useful for showing the true shape of the cross-section of some elongated object, such as a thread rod (螺杆), shaft (轴), sleeve (套管), arm, spoke (轮辐), or rib (肋) as shown in Fig. 7.32.

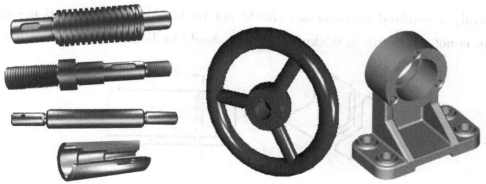

Fig. 7.32 Some elongated objects.

153

As shown in Fig. 7.33, to obtain such a revolved cross-section, an imaginary cutting-plane is passed through the member perpendicular to the longitudinal axis and then revolved 90° to bring the resulting view into the front view.

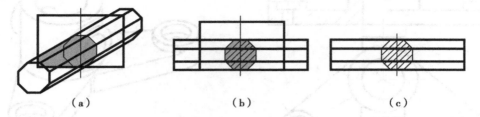

Fig. 7.33 Forming of a revolved cross-section.

The superimposition of the revolved cross-section does not require the removal of all original lines covered by it. In other words when the revolved cross-section is revolved 90° to be placed in the views, the thick lines in the view are also thick lines, so in Fig. 7.34, figure (a) is correct while figure (b) incorrect.

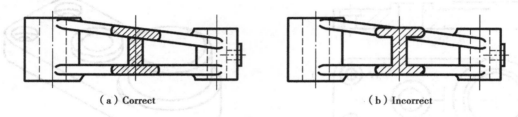

Fig. 7.34 Drawing of revolved cross-sections.

Other examples of revolved cross-sections are shown in Fig. 7.35. Note that when a revolved cross-section is broken, there is no wave line at breaking as shown in Fig. 7.35 (b).

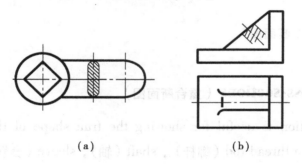

Fig. 7.35 Examples

In general, a revolved cross-section should not be labeled. However, if the shape of the cross-section is not symmetric in thickness, arrows should be labeled as shown in Fig. 7.36.

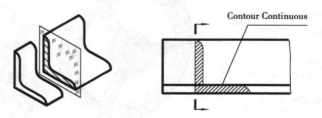

Fig. 7.36 Anisomerous (不对称的) revolved cross-section.

7.3.2 Removed cross-section

A removed cross-section is similar to a revolved cross-section, except that it does not appear on a view but instead is drawn out of the view and appears adjacent to it.

In general, removed cross-sections should be labeled. The cutting-plane lines are drawn in the view from place where the object is cut. At the ends of the cutting-plane lines, the same two capital letters are written. Correspondingly, the removed cross-section is named by the same two capital letters separated by a dash. Such cross-section is read as cross-section *A-A* or cross-section *B-B*. Besides, cross-section *B-B* or cross-section *D-D* may be arranged at the extended line of the cutting-plane lines and also be arranged according to the position of a primary view, but the rotating symbol ⌒ must be added as shown in Fig. 7.37 (a).

In order to represent the object clearly, except the front view is a broken-out section, four cross-sections and a partial view, view *E* are adopted. Fig. 7.37 (b) is its pictorial drawing.

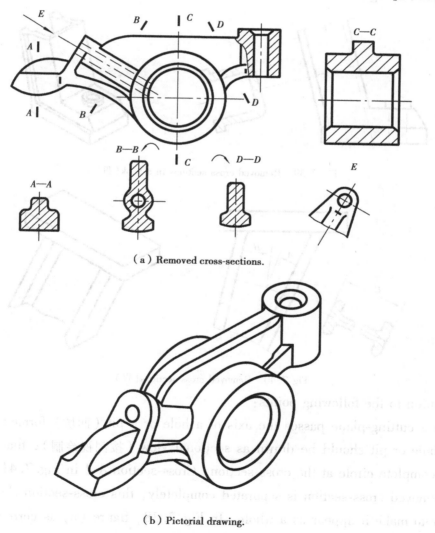

(a) Removed cross-sections.

(b) Pictorial drawing.

Fig. 7.37 Removed cross-sections(I).

Note that if the shape of the cross-section is unsymmetric in thickness, arrows should be labeled as shown Fig. 7.38.

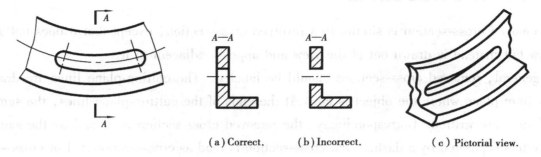

Fig. 7.38 Removed cross-sections(Ⅱ).

Sometimes it is convenient to place removed cross-sections on the extended line of the cutting-plane line as shown in Fig. 7.39. If a removed cross-section is obtained by intersecting cutting planes, it is generally broken for the sake of the convenience in drawing. However, dimensions should be agreed with relation C < (A + B) as shown in Fig. 7.40.

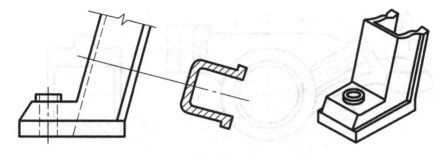

Fig. 7.39 Removed cross-sections in a hook(Ⅲ).

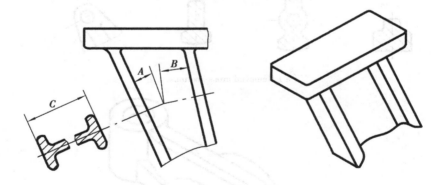

Fig. 7.40 Removed cross-sections(Ⅳ).

Pay attention to the following points:

1) When a cutting-plane passes the axis of a hole or a pit (凹坑) formed by revolutionary surface, the hole or pit should be drawn as sectional view (按剖视绘制), that is, the cylinder appears as a complete circle at the cross-section. Cross-section A-A in Fig. 7.41 (a) is correct.

2) If a removed cross-section is separated completely, this cross-section should be drawn as sectional view to make it appear as a whole. In Fig. 7.38, figure (a) is correct but figure (b) incorrect. Fig. 7.38 (c) is the pictorial drawing of the object.

Lastly, notice the difference between cross-section and sectional views. In cross-section, only the current cut area is drawn, while section views include all visible features of remained part

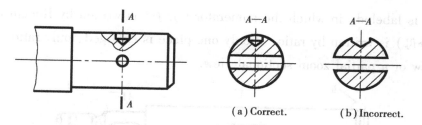

Fig. 7.41 Removed cross-sections.

of the object as shown in Fig. 7.42.

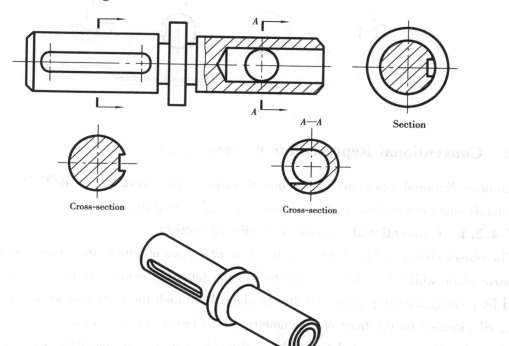

Fig. 7.42 Difference between cross-sections and sectional views.

7.4 Other Representation Methods（其他表达方法）

From previous studying, it is known to represent the outside shape of an object with views; to express the inside shape of an object with sectional view; to represent the cross shape of some elongated object such as shaft, arm, with cross-section. In engineering practice, there are some other representation methods used to represent object, such as a partial zoom view（局部放大图） and a series of conventional representation.

7.4.1 Partial zoom view（局部放大图）

A partial zoom view is one view in which certain structure of an object is drawn by an enlarged scale. It may be a partial view or a partial sectional view as shown in Fig. 7.43. In front view, the places where small structures will be enlarged are encircled（环绕） by thin circle and numbered by Roman numbers. Above a partial zoom view or a partial zoom sectional view, a

157

fraction (分数) is labeled, in which the numerator (分子) is written by Roman number, and the denominator (分母) is written by ratio. If only one place is enlarged, only ratio is written above a partial zoom view or a partial zoom sectional view.

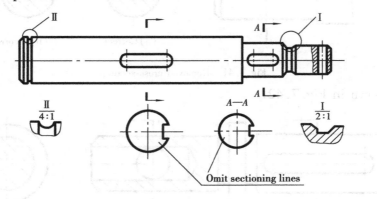

Fig. 7.43 A zoom view.

7.4.2 Conventional Representation (惯例表达法)

Chinese National Standard of Technical Engineering Drawing (GB/T 16675.2—1996) recommends some conventional practices used in engineering drawing.

7.4.2.1 Conventional treatment of ribs in section

The object shown in Fig. 7.44 (a) has four ribs, two of which are located on the front-rear symmetric plane while the others are located on the left-right symmetric plane. The front view is formed by passing a cutting-plane parallel to plane V through the front-rear symmetric plane of the object, ribs located on the front-rear symmetric plane being cut longitudinally (纵向). Under the condition, the ribs are regarded to not be sectioned as if the cutting-plane were just in front of them, that is to say, the section lines are eliminated (删除) from the ribs. A true sectional view of the ribs with section lines givens a misleading effect suggesting a cone shape (Fig. 7.44 (b)). However, in Fig. 7.44 (c), section A-A is formed by passing a cutting-plane parallel to plane H, through the object, four ribs being cut transversely (横向). Under the condition, ribs are always sectioned.

Another example about ribs is the shaft bracket shown in Fig. 7.45. The left view of the shaft bracket is drawn as a full section in which the cutting-plane is parallel to plane W and it passes through the axis of the cylinder sleeve. At the same time, the rib in the left-right symmetrical plane is cut longitudinally, so the rib is not sectioned. However, the rib in the front-rear symmetrical plane is cut transversely, so the rib is sectioned. The top view of the shaft bracket is full section A-A in which two ribs are cut transversely by cutting-plane A-A, thus they are sectioned.

Chapter 7 Sectional Views and Cross-Sectional Views

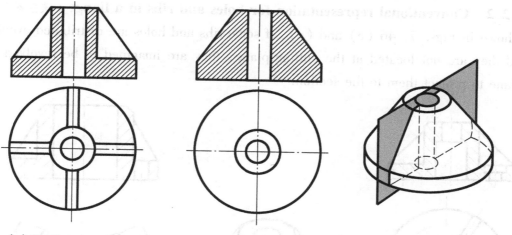

(a) The rib is cut longitudinally.

(b) The rib is misled.

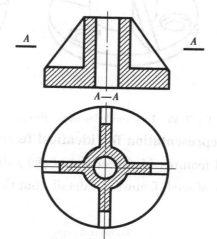

(c) The rib is cut transversely.

Fig. 7.44 Ribs in section.

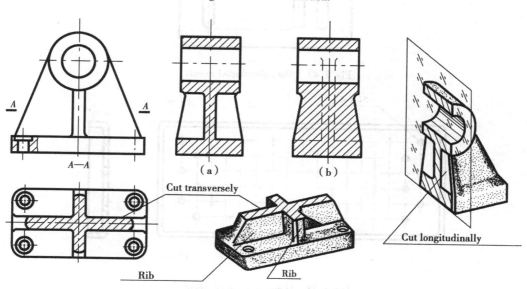

Fig. 7.45 Shaft bracket.

7.4.2.2 Conventional representation for holes and ribs in a flange (法兰盘)

As shown in Figs. 7.46 (a) and (c), if some ribs and holes are distributed evenly on the flange and they are not located at the cutting-plane, they are imagined to be revolved onto the cutting-plane to project them to the section.

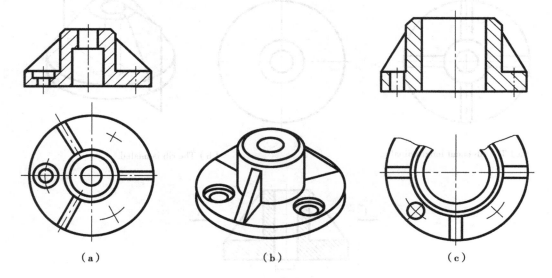

Fig. 7.46 Ribs and holes on the flanges.

7.4.2.3 Conventional representation for identical features (同样结构)

When a number of identical features (holes or slots (槽)) distribute regularly on the object, it is only necessary to draw a few of such features in detail, but the numbers must be noted (Figs. 7.47 and 7.48).

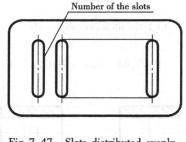

Fig. 7.47 Slots distributed evenly.

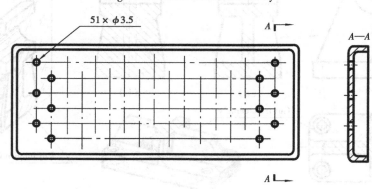

Fig. 7.48 Holes distributed evenly.

7.4.2.4 Conventional representation for some figures and intersection lines

① Sectional lines on removed cross-section are allowed to omit if this will not lead to misunderstanding (Fig. 7.43).

② A planar symbol (two intersecting thin lines) is applied to indicate a small plane (Fig. 7.49).

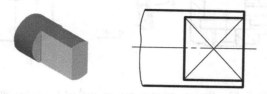

Fig. 7.49 Use planar symbol to express plane.

③ Intersections and interim lines (过渡线) can be replaced by circles or straight line if this will not lead to misunderstanding (Fig. 7.50).

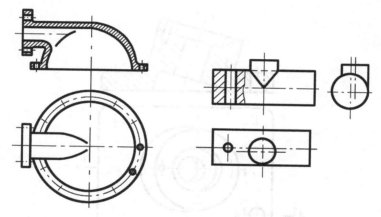

Fig. 7.50 Intersections and interim lines in conventional practice.

④ If a simple feature located in front of the cutting-plane is expressed in a section, draw it by phantom lines (假想线) as shown in Fig. 7.51.

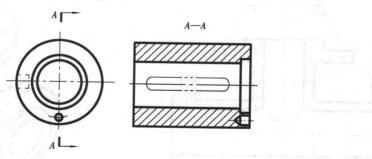

Fig. 7.51 Feature ahead the cutting plane.

⑤ Broken-out section can be applied to sectional view and, in which, section lines should be drawn as shown in Fig. 7.52 and label it above the leader (指引线).

⑥ Circles or circular arcs with an angle smaller than 30° can be drawn as circles or circular arcs (Fig. 7.53).

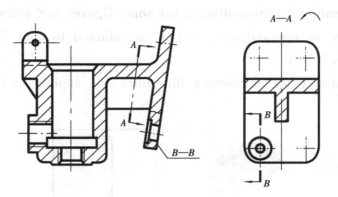

Fig. 7.52 Broken-out view can be applied to section view.

⑦ Holes distributed evenly in a flange (法兰盘) can be drawn as shown in Fig. 7.54. At upper flange, a partial top view is drawn a half only. Similarly, at left flange, a partial left view is drawn a half only. Actually each flange has four small holes of equal diameters.

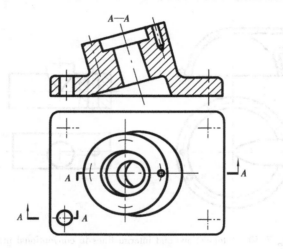

Fig. 7.53 Circles or arcs with an angle smaller than 30° with projection plane H.

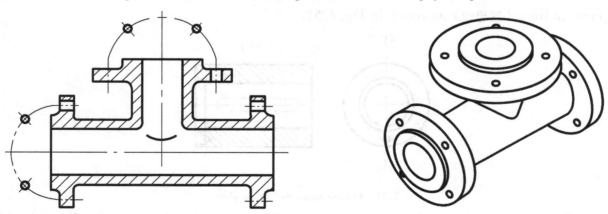

Fig. 7.54 Holes distributed symmetrically in the flange.

⑧ Partial view of a symmetrical structure can be drawn and arranged nearby the structure. In Fig. 7.55 (a), the top view of the key seat in a shaft is arranged above the front view of the shaft. In Fig. 7.55 (b), the top view of the key seat in a cylinder sleeve is also arranged above the front view of the object

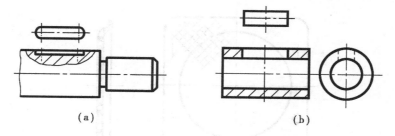

Fig. 7.55 Partial view of symmetrical structure.

7.4.2.5 Conventional representation for small features

1) Small features expressed clearly on one view can be omitted or simplified on other views. In Fig. 7.56, two intersections located on the upper and lower are omitted as shown in Fig. 7.56 (a) while four intersections are simplified into straight lines in Fig. 7.56(b). In Fig. 7.56 (c) is the pictorial drawing of the object represented by Fig. 7.56(a).

2) If it does not cause misunderstanding, the small fillet (小圆角), small obliquity (小倾角) and small chamfer (小倒角) can be omitted on the views, but the dimensions must be indicated (Fig. 7.57).

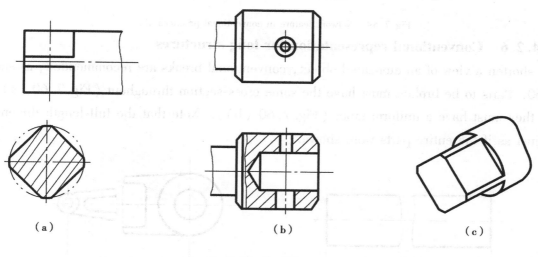

Fig. 7.56 Small features.

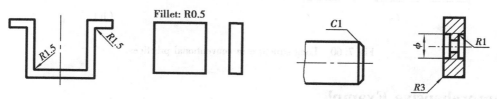

Fig. 7.57 Small features.

3) Network feature (网纹) can be expressed by thin lines as shown in Figs. 7.58 and 7.59.

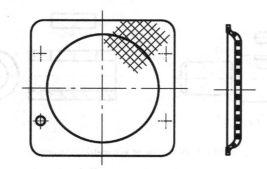

Fig. 7.58 Network feature in conventional practices Ⅰ.

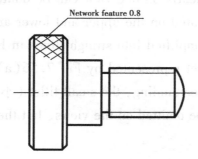

Fig. 7.59 Network feature in conventional practices Ⅱ.

7.4.2.6 Conventional representation for long structures

To shorten a view of an elongated object, conventional breaks are recommended, as shown in Fig. 7.60. Parts to be broken must have the same cross-section throughout (Fig. 7.60 (a)) or if tapered they must have a uniform taper (Fig. 7.60 (b)). Note that the full-length dimension is given, just as if the entire parts were shown.

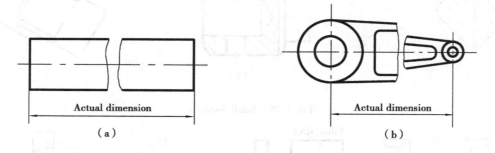

Fig. 7.60 Long structure in conventional practices.

7.5 Comprehensive Example

To express an object, it is necessary to design more then one expression scheme by using views, sectional views or cross-section and so on. The fundamentals of an adequate expression scheme are as follows. First the views are easy to read, and secondly views should be as concise (简洁的) as possible, and thirdly the structure and shape are represented fully and clearly.

Fig. 7.61 shows a pipe joint (管接头) and its expression scheme is as follows.

Step 1　Analysis.

The joint is composed of one vertical pipe and two horizontal pipes in which at the ends of the pipes are designed with flanges. Left horizontal pipe is supported by a rib (Fig. 7.61 (a)).

Step 2 Choose the projection direction for the front view.

Usually, a projection direction that reflects shape features mostly is selected as the projection direction of the front view. To give clear interior structures, two intersecting cutting-planes in which one is parallel to plane V and the other one is perpendicular to the plane H and inclined plane V are supposed to pass through the axes of two horizontal pipe holes. So, the front view is an aligned full section A-A.

Step 3 Choose other views.

Two cutting-planes parallel to plane H are supposed to pass the axes of the two horizontal pipe holes, so the top view obtained is an offset full section B-B (Fig. 7.61 (b)). In addition, sectional view C-C and D-D are used to describe the shape of flanges.

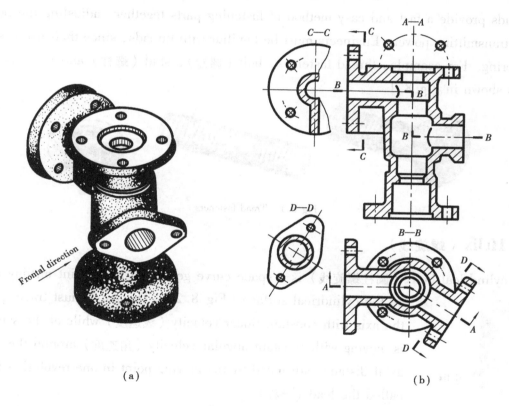

(a)   (b)

Fig. 7.61 A pipe joint.

From this example, the steps for expressing an object should be as follows.

1) Being familiar with the object's form and feature;

2) Choosing the projection direction that reflects shape features mostly as the projection direction of front view.

3) Choosing other views.

4) Deciding the scheme to express an object in the best way.

Chapter 8　Threads, Fasteners and Gears

In engineering, there is always the necessity for assembling parts either with permanent fastenings (永久连接) such as rivets (铆钉) or knockdown fastenings (可拆卸的连接) such as bolt joining. This chapter is focuses on the knockdown fastenings including threads, threaded fasteners, some other fasteners. Besides, gears will be introduced simply in this chapter.

8.1　Thread (螺纹)

Threads provide a fast and easy method of fastening parts together, adjusting the position of parts and transmitting power. Engineers must be familiar with threads, since their use is so widely in engineering. For example, thread fasteners, bolt (螺栓), stud (螺柱) and nut (螺母) have threads as shown in Fig. 8.1.

Fig. 8.1　Tread fasteners

8.1.1　Helix (螺旋线)

The cylindrical helix (圆柱螺旋线) is a space curve generated by a point moving uniformly on the cylindrical surface (Fig. 8.2). The point must travel parallel to the axis with constant linear velocity (线速度) while at the same time it is moving with constant angular velocity (角速度) around the axis. The axial distance advanced by the moving point in one revolution (一圈) is called the lead (导程).

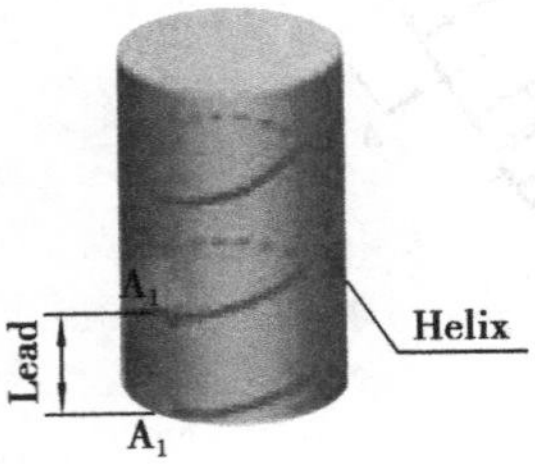

Fig. 8.2　Cylindrical helix.

Two views of the cylindrical helix are drawn as follows. The helix's top view is circumference while its front view needs to be made.

As shown in Fig. 8.3 (b), the circle is divided into any number of equal parts, such as 12 parts. Number the divisions from 1 to 12. On the front view of the cylinder, set off the lead and divide it into uniform parts as many as that of the circle. Number the divisions from 1_0 to 12_0. Through each division, draw a line perpendicular to the axis of the cylinder, which is called division line. When the generating point has rotated one-twelfth around the cylinder, it has also advanced one-twelfth of the lead; when halfway around the cylinder, it will have advanced one-half the lead; and so on. Thus, points on the front view of the helix can be found by projecting points in the circular view to the division lines in the front view. For example, from point 1, draw projection line to the front view to intersect division line 1_0 at $1'$. By the same way, points from $2'$ to $12'$ can be obtained. Points from $1'$ to $6'$ are connected smoothly with a thick line while points from $6'$ to $12'$ are connected smoothly with hidden line.

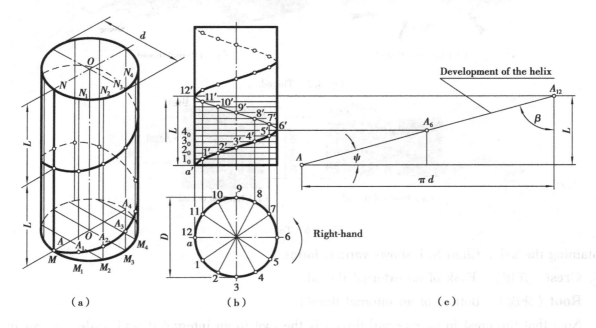

Fig. 8.3 Helix.

If the cylinder is developed (展开), the helix will appear as a straight line inclined to the base at an angle known as the helix angle (螺旋角), whose tangent is $L/\pi d$, where L is the lead and d the diameter (Fig. 8.3 (c)).

The helix is base to form thread (螺纹).

8.1.2 Thread and thread terms (螺纹术语)

A threaded part (螺旋体) is formed by a planer figure like triangle or rectangle moving along helix on a cylinder or cone as shown Fig. 8.4.

The threaded part is known as thread simply. The thread is cut as the tool (刀具) travels parallel to the axis of the revolving work (工件) at a constant speed. The thread machined (加工) on external surface of the revolution is called external thread (Fig. 8.5 (a)). The thread machined in internal surface of the revolution is called internal thread (Fig. 8.5 (b)).

Understanding threads begins with learning their terms (Fig. 8.6).

Fig. 8.4 A threaded part.

Tooth (牙) Convex (凸起) part on a thread.

Groove (槽) Concave (凹入) part on a thread.

Form of thread (牙形) The shape of the cross-section (断面) of thread cut by a plane

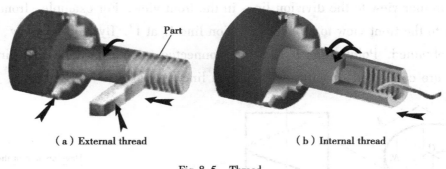

(a) External thread (b) Internal thread

Fig. 8. 5 Thread.

(a) External thread (b) Internal thread

Fig. 8. 6 Terms.

containing the axis. Chart 8. 1 shows various forms.

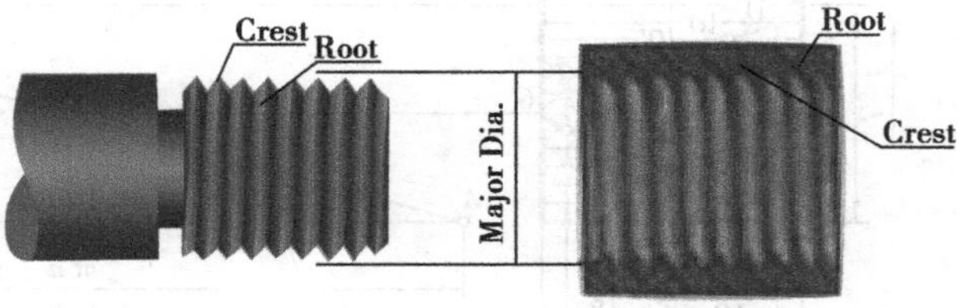

Crest (牙顶) Peak of an external thread.

Root (牙底) Bottom of an internal thread.

Note that the crest in an external thread is the root in an internal thread while the root in an external thread is the crest in an internal thread.

The vertical distance between crest and root is called depth of thread.

Major diameter (螺纹大径 d, D) The largest diameter on an internal or external thread. It is also called nominal diameter (公称直径) except pipe thread. It is denoted by letter d for external thread but capital letter D for internal thread.

Minor diameter (螺纹小径 d_1, D_1) The smallest diameter on an internal or external thread. It is denoted by letter d_1 for external thread but capital letter D_1 for internal thread. The values of minor diameters of diversified threads are lised in the Appendix 2. In drawing, the minor diameter is approximately 0. 85 times the major diameter, that is, $d_1 \approx 0.85d$ for internal thread, $D_1 \approx 0.85D$ for external thread.

Pitch diameter (中径) The diameter of an imaginary cylinder whose generatrix is located a place where the width of the tooth equals that of groove. The imaginary cylinder is called pitch diameter cylinder and its generatrix is called the pitch diameter line (中径线).

Pitch (螺距) The distance between corresponding points on two adjacent teeth of the thread measured along the pitch diameter line (Fig. 8. 7 (a)).

Lead (导程) The distance a threaded part moves axially, with respect to a fixed mating part, in one complete revolution (Fig. 8. 7 (b)).

Single thread (单线螺纹) A threaded part having only one helix on the cylinder. (Fig. 8. 7 (a)).

Multiple thread (多线螺纹) A threaded part having two or more helices on the cylinder.

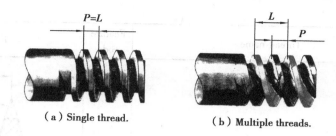

(a) Single thread.   (b) Multiple threads.

Fig. 8.7 Single and double threads.

(Fig. 8.7(b))

In general, a thread is a single thread unless otherwise specified. On a single thread, the lead and pitch are equivalent. For a multiple thread, the lead is an integral multiple of the pitch; that is, on a double thread, lead is twice the pitch; on a triple thread (三线螺纹), lead is three times the pitch. A multiple thread permits a more rapid advance than a single thread when turned 360°.

Right-hand thread (右旋螺纹) A right-hand threaded part advances into a thread hole when turned clockwise (顺时针方向). A right-hand external thread slopes upward to the right when its axis is vertical (Fig. 8.8 (b)).

Left-hand thread (左旋螺纹) A left-hand threaded part advances into a thread hole when turned counterclockwise (反时针方向). A left-hand external thread slopes upward to the left when its axis is vertical (Fig. 8.8 (a)). The left-hand thread is designated by *LH*.

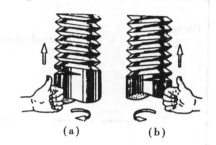

(a)　(b)

Fig. 8.8 Rotating direction of thread.

Among above terms, the form of thread, major diameter and pitch are basic elements. If the basic elements are designed by Chinese National Standard for Technical Drawing, they are standardized threads and their structure and size can be found in Table 1 to Table 5 in Appendix 2 at the end of the book or manufacture catalogue. In engineering practice, external thread engages with (旋合) internal thread to work. Which types of external and internal threads can be engaged with each other? Only if all five elements including the form of thread, major diameter, pitch, numbers of threads and rotating direction of thread are identical, the external and internal threads can be engaged with each other.

Chart 8.1 Thread types and feature codes

Usage	Thread name		Feature code	Form of thread
Fastening (紧固)	Metric threads (普通螺纹)	Coarse thread (粗牙普通螺纹)	M	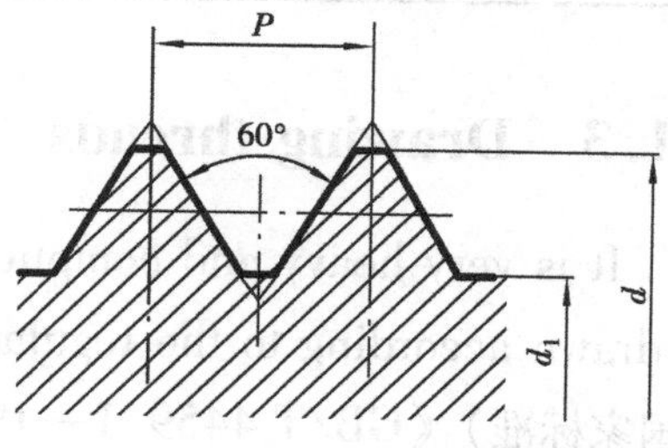 (a) Metric thread form (M)
		Fine thread (细牙普通螺纹)		

169

(Continued)

Usage	Thread name		Feature code	Form of thread
Transmission (传输)	Trapezoid thread (梯形螺纹)		Tr	(b) Trapezoid thread form (Tr)
	Sawtooth thread (锯齿形螺纹)		B	(c) Sawtooth thread form (B)
	Square thread (矩形螺纹)			(d) Square thread form
Coupling (连接)	straight pipe threads (非螺纹密封的管螺纹)	Cylindrical external thread (圆柱外螺纹)	G	
		Cylindrical internal thread (圆柱内螺纹)	G	
	Taper pipe threads (螺纹密封的管螺纹)	Taper external thread (圆锥外螺纹)	R	
		Taper internal thread (圆锥内螺纹)	R_c	
		Cylindrical internal thread (圆柱内螺纹)	R_p	

8.1.3 Drawing threads

It is very heavy and complicated work to draw thread in actual projection. So, thread should be drawn according to the regulations of Chinese National Standard of Technical Drawing (技术制图国家标准)《GB/T 4459.1—1995》.

8.1.3.1 Regulated drawing of an external thread（外螺纹的规定画法）

As shown in Fig. 8.9, in the front view, the major diameter lines are drawn with thick line while the minor diameter lines are drawn with thin line and extended to the chamfer（倒角）. Where the thread is ended, a thick line is drawn, which is called thread ended line（螺纹终止线）. Note that in the section view of an external thread, thread ended line is drawn as a short segment AB from the crest to the root (Fig. 8.10).

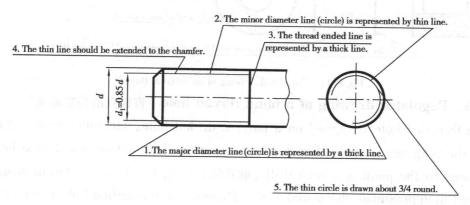

Fig. 8.9 Stipulated drawing of external thread.

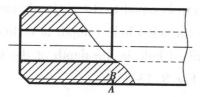

Fig. 8.10 Stipulated drawing of external thread.

In the left view, the major diameter circle is drawn with thick line. Minor diameter circle is represented by partial thin circle and it is as long as 3/4 circumference. Chamfer circle is omitted in the left view.

8.1.3.2 Regulated drawing of an internal thread（内螺纹的规定画法）

Normally, internal thread is represented by a section in the front view.

The section view of internal thread is shown in Fig. 8.11. The major diameter lines are drawn with thin lines while the minor diameter lines with thick lines, which are not extended into the chamfer. Thread ended line is drawn with a thick line. Note that in the section view of internal thread, thread ended line is drawn from the root on one side to the root on the other side as shown in Fig. 8.11, the thread ended line CD. Section lines are drawn with thin lines and let them traverse the minor diameter lines to reach the major diameter lines.

In the left view, the major diameter circle is represented by partial thin circle as long as 3/4 circumferences only. The minor diameter circle is drawn with thick circle. Chamfer circle is omitted. Note that in the left view of Fig. 8.11, the largest circle is the projection of the outside cylindrical surface.

Fundamentals of Engineering Drawing

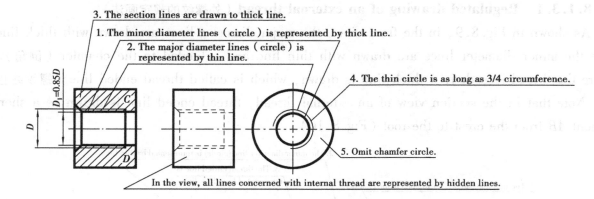

Fig. 8.11 Stipulated drawing of an internal thread.

8.1.3.3 Regulated drawing of a blind thread hole (盲螺孔的规定画法)

Before a threaded hole is tapped on a part, a drilled hole, clearance hole (光孔) must be drilled to let the plug tap (丝锥) (Fig. 8.13 (a)) enter. The bottom of a drilled hole is conical point, as formed by the point of a twist drill (麻花钻) (Fig. 8.12 (a)). On drawings, an angle of 120° is used to approximate the actual 118°. Diameter of the drilled hole is equal to the minor diameter of the threaded hole D_1 (Fig. 8.12). A plug tap is used to tap threaded hole in a drilled hole and the diameter of the plug tap is equal to the major diameter of the thread, D,. Normally, the drilled hole is deeper than the threaded hole (depth of drilled hole = depth of threaded hole + $0.5D$). Note that the depth of drilled hole is the depth of the cylindrical portion of the hole but does not include the cone point (Fig. 8.13).

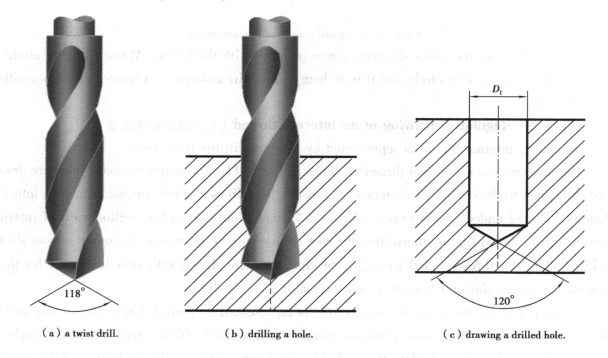

(a) a twist drill.　　(b) drilling a hole.　　(c) drawing a drilled hole.

Fig. 8.12 A drilled hole.

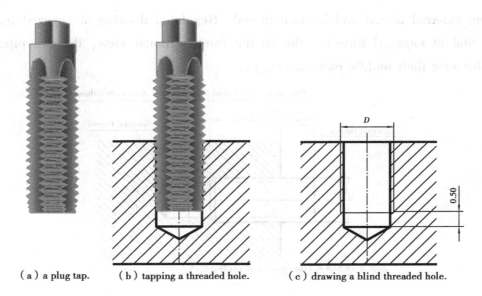

(a) a plug tap.　　(b) tapping a threaded hole.　　(c) drawing a blind threaded hole.

Fig. 8.13　A blind threaded hole.

8.1.3.4　Regulated drawing of intersecting thread holes（螺孔交线的规定画法）

No matter a threaded hole intersects with a drilled hole (Fig. 8.14(a)), or a threaded hole (Fig. 8.14 (b)), only the intersections of minor diameter lines are drawn with a thick line while the intersection of major diameter lines are omitted.

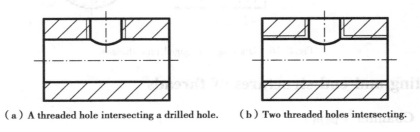

(a) A threaded hole intersecting a drilled hole.　　(b) Two threaded holes intersecting.

Fig. 8.14　Intersections.

8.1.3.5　Regulated drawing of engaged threads（旋合螺纹的规定画法）

When an external thread and an internal are engaged together, in order to represent their inner structures, usually they are sectioned by a cutting-plane passing through their axis. As shown in Fig. 8.15, two engaged objects with thread are cut by a frontal plane passing through the axial plane of threaded rod（螺杆）. In the section, the engaged part is drawn as external thread and the non-engaged parts are drawn as the forms of external thread and internal thread, respectively. Especially note that thick lines and thin lines must be alignment（平齐）because they represent respectively major and minor diameters of the external thread and internal thread.

In section of engaged thread, it is customary for threaded rod not to be sectioned. For example, as shown in Fig. 8.15, the threaded rod is cut actually, but it is not sectioned in drawing. Note that the section lines should traverse（穿过）the thin lines and reach the thick lines. In other words, the section lines are interdicted（阻断）at thick line.

Fig. 8.16 illustrates two pipe threads（管螺纹）engaged together. The left end is a pipe with external thread while right end is a pipe with internal thread. The two pipes are coupled（连接）

by engaging external thread and internal thread. Regulated drawing of engaged pipe threads is similar to that of engaged threads. But in the front sectional view, the two pipe threads are sectioned because their middle parts are empty.

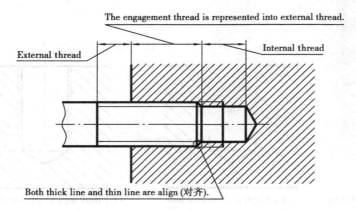

Fig. 8.15 Drawing of engaged threads.

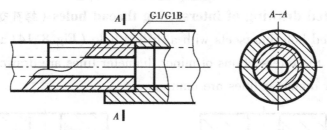

Fig. 8.16 Drawing of engaged pipe threads.

8.1.4 Starting and end structures of threads

8.1.4.1 Chamfer (倒角)

When an external thread engages with an internal thread, in order to assemble them easily and avoid damaging the ends, chamfers are made on their ends. Fig. 8.17 (a) is the chamfer on external thread while Fig. 8.17 (b) is the chamfer on internal thread. Chamfered values may be checked by Table 7 in Appendix 2.

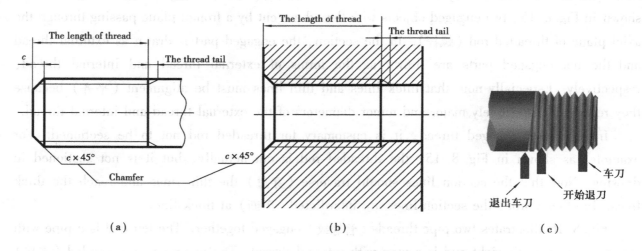

Fig. 8.17 Chamfers and thread tails.

8.1.4.2 Thread tail (螺尾)

At the end of a thread, a segment of shallower (浅的) thread formed due to the withdrawing of machining tool (退刀) is called thread tail (Fig. 8.17 (c)). The thread tails of the external and internal threads are shown in Fig. 8.17 (a) and (b).

8.1.4.3 Tool escape (退刀槽)

In order to avoid the thread tail, before cutting thread, a groove called tool escape is made. The external thread's tool escape is shown in Figs. 8.18 (a) and (b), in which the diameter of the tool escape is smaller than minor diameter of the external thread. The internal thread's tool escape is shown in Figs. 8.18 (c) and (d), in which the diameter of the tool escape is larger than major diameter of the internal thread.

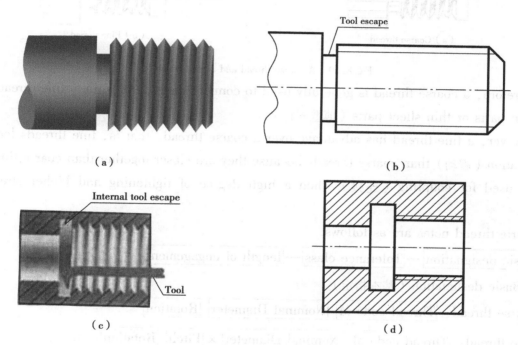

Fig. 8.18 Thread tool escape.

8.1.5 Thread notes (螺纹标记)

Drawings of threads are only symbolic representations and are inadequate to give the details of a thread unless accompanied by notes.

8.1.5.1 Metric thread notes (普通螺纹的标记)

Metric threads include coarse threads and fine threads. For a given metric thread in nominal diameter, a coarse thread has the largest pitch among all pitches.

Example

Coarse thread, $M = 36$, $P = 5$;

Fine thread, $M = 36$, $P = 3, 2$ or 1.5.

Above data are listed in Appendix 2 Table 1.

This means that a coarse thread permits more rapid advance than a fine thread when turned 360°. Furthermore, for a given metric thread in nominal diameter, the minor diameter of a coarse thread is smaller than that of any fine thread.

Example

Coars thread, $M=36$, $d_1=31.62$;

Fine thread, $M=36$, $d_1=32.752$, 33.835 or 35.376.

Above data are listed in Appendix 2 Table 1.

This means that tooth of a coarse thread is larger than that of any fine thread (Fig. 8.19).

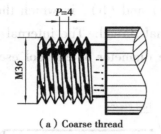 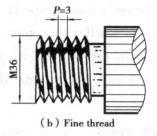

(a) Coarse thread   (b) Fine thread

Fig. 8.19 A coarse thread and a fine thread.

Therefore, a coarse thread is generally used to connect large parts while a fine thread is used in presion parts or thin sheet parts (薄壁零件).

However, a fine thread has advantage over a coarse thread, that is, fine threads loosen less with vibration (震动) than coarse threads because they are closer together than coarse threads, so they are used for coupling (连接) when a high degree of tightening and higher strength are required.

Metric thread notes are as follows.

| Basic designation |—| Tolerance class |—| length of engagement |

1) Basic designation (基本指定)

Coarse thread: | Thread code M | | Nominal Diameter | | Rotation |

Fine thread: | Thread code M | | Nominal Diameter | × | Pitch | | Rotation |

Note that threads can be classified as those of right-hand thread and left-hand thread. Letters LH indicates left-hand thread and RL indicates right-hand thread. In basic designation, letters RH is omitted. while LH must be written. As stated previously, a left-hand thread means that the thread will be tightened by going anticlockwise. In general, threads are right-hand threads except that a right-hand thread might loosen under vibration or load.

Examples

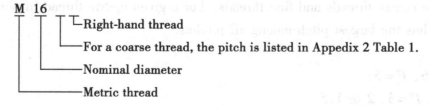

176

2) Tolerance class (公差等级)

The tolerance class includes the symbol for the pitch-diameter tolerance followed by the symbol for the crest-diameter (顶径) tolerance. Each of these symbols consists of a number indicating the tolerance grade followed by a letter indicating the tolerance position. The tolerance class is separated from the basic designation by a dash.

Example

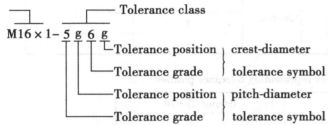

When the pitch-diameter and crest diameter tolerance symbols are identical, the symbol is only given once. It is not repeated.

Example

M16 × 1-6g
└─Pitch diamtter and crest-diameter tolerance symbols (equal)

Note 1. Crest-diameter designate the major diameter in an external thread and the minor diameter in an internal thread.

Note 2. In tolerance class, the capital letter indicating tolerance position indicates an internal thread while the small letter indicating tolerance position indicates an external thread.

Example

M16 × 1-6H
       │└─tolerance position of an internal thread
       └─tolerance grade
       ┌─Pitch-diameter and crest-diameter tolerance class

Tolerance grades are a series of numbers representing the accuracy of manufacturing. For a given tolerance grade number, the purpose is to give the same level of accuracy in manufacturing thread. Available grade numbers are listed in Chart 8.2.

Tolerance increases as the tolerance-grade numbers increase, but manufacturing accuracy decreases. For example, Grade 4 means a small tolerance but high accuracy in manufacturing. Grade 6 means a medium, and grade 8 means a large tolerance but low accuracy in manufacturing. In general, grades of less than 6 are best suited to fine series while grades greater than 6 are best suited to coarse series.

Tolerance position is the position of the tolerance zone relative to the basic size (check Chapter 9 for in-depth information). It expresses an allowance in engaging thread. The smaller the allowance is, the tighter the engaging thread is.

In chart 8.3, the letter e permits a large allowance; g, G express a small allowance; and h, H indicate a zero allowance.

Chart 8.2 Tolerance Grades

External Thread			Internal Thread	
crest diameter (Major Diameter)	Pitch Diameter		crest diameter (Minor Diameter)	Pitch Diameter
—	3	Fine	—	—
4	4		4	4
—	5		5	5
6	6		6	6
—	7	Coarse	7	7
8	8		8	8
—	9		—	—

Chart 8.3 Letters representing tolerance position

External Thread	Internal Thread
e = large allowance	
g = Small allowance	G = Small allowance
h = No allowance	H = No allowance

3) Engagement length

There are three engagement lengths available: normal (N), short (S), and long (L). Most applications use normal lengths and symbol N is omitted. The length of thread engagement is added to the end of the tolerance class designation and separated by a dash.

For example, M16 × 1-6g-L
　　　　　　　　　　└─ Long length of engagement

8.1.5.2 Thread notes for trapezoid and sawtooth threads

Thread notes of trapezoid and sawtooth threads are identical to metric thread, ie

| Basic designation |—| Tolerance class |—| Length of engagement |

1) Basic designation

Trapezoid and sawtooth threads can be classified into single thread and multiple threads. Their basic designations are as follows.

Single thread:

| Thread code Tr or B | Nominal Diameter | × | Pitch | Rotation |

Multiple threads:

| Thread code Tr or B | Nominal Diameter | × | Lead (P) | Rotation |

Rotation direction is the same with the metric thread, ie "RH" is omitted while "LH" must be Written.

2) Tolerance class

The tolerance class of trapezoid and sawtooth threads include only the pitch-diameter tolerance class.

3) Engagement length

The engagement length of trapezoid and sawtooth threads is the same with the metric thread.

Example

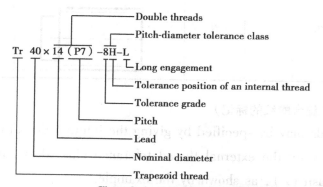

Chart 8.4 Labeling threads for metric, trapezoid and sawtooth threads

Types	Feature code	Nominal diameter	lead	Pitch	Thread numbers	Rotation direction	Tolerance range code		Engagement length	Labeling example
							Pitch diameter	Major diameter		
Coarse thread	M	16	2	2	1	Right	8h	8h	N	M16-8b
			2	2	1	Right	7h	7h	N	M16-7H
Fine thread	M	36	1	1	1	Left	5g	6g	L	M36×1LH-5g6g-L
Trapezoid thread	Tr	36	6	6	1	Left	7e		L	Tr36×6LH-7e-LH
		40	14	7	2	Right	8H		N	Tr40×14(P7)8H

(Continued)

Types	Feature code	Nominal diameter	lead	Pitch	Thread numbers	Rotation direction	Tolerance range code		Engagement length	Labeling example
							Pitch diameter	Major diameter		
Sawtooth thread	B	60	16	8	2	Left	7e		N	B60×16(p8)-7c

8.1.5.3 Engaging threads notes (旋合螺纹的标记)

A required fit between engaging threads may by specified by giving the internal thread tolerance class designation followed immediately by the external thread tolerance class designation. The two designations are separated by a slash (/), as shown by the example.

Example, M16×1-5H/5g6g

8.1.5.4 Pipe thread notes (管螺纹的标记)

Straight pipe threads and taper pipe threads are two general types of pipe threads standardized by Chinese National Standard of Technology Drawing. Straight pipe threads are used for pressure-tight joints for pipe couplings. Note that before an external pipe thread engages with an internal thread pipe thread, sealer such as strip of waterproofer (防水带) is wrapped round the external pipe thread. Taper pipe threads are used to provide a metal-to-metal joint, eliminating the need for a sealer, and are used in refrigeration (制冷), marine (海运), automotive, aircraft, and ordnance work (军火).

Pipe thread note begins with letter G, R, R_C or R_P followed by the nominal size in inches and LH separated by a dash. Absence of LH indicates a RH thread. Examples are as follows:

"G1" means right-hand straight pipe thread of which nominal size is 1 inch.

R3/4-LH means left-hand direction taper pipe thread of which nominal size is 3/4 inch.

Note that for straight pipe threads, either cylindrical external thread or cylindrical internal thread, the code of form of thread is letter G. In order to identify them from thread notes, pitch diameter tolerance grade A or B is added at the end of the nominal size of the cylindrical external thread. Tolerance grade A is higher than tolerance grade B in accuracy. Examples are as follows:

G1A means right-hand cylindrical external thread of which nominal size is 1 inch.

G1 means right-hand cylindrical internal thread of which nominal size is 1 inch.

Lastly, it should be pointed out that the nominal size of the pipe thread indicates the diameter of hole (孔径) rather than major diameter. In other words, from pipe thread note, the major diameter is not known. Major diameters of all pipe threads are listed in Tables 4 and 5 in Appendix 2. For example, from "G1A", it is checked by Table 4 that the nominal size is 1 inch or 25.4 mm, however the major diameter of the thread is 33.249 mm.

After writing pipe thread, it is labeled on the drawings. First, draw a leader line from pipe thread's major diameter, then write pipe thread notes on the leader line.

Diversified labeling examples are shown in Chart 8.5.

Chart 8.5 Labeling pipe threads

Types	Feature code	Nominal diameter	Rotation direction	Tolerance Grade code	Labeling example
Straight pipe thread	G	1	Right	A	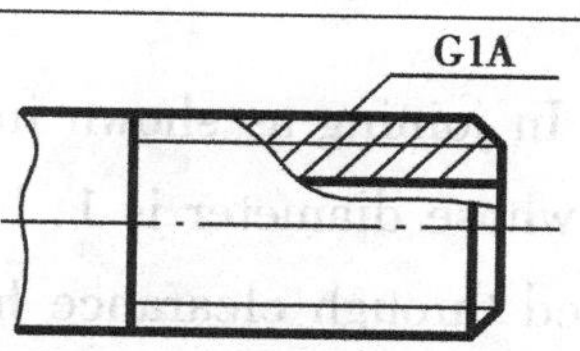 G1A
Taper pipe thread	Rc	314	Left		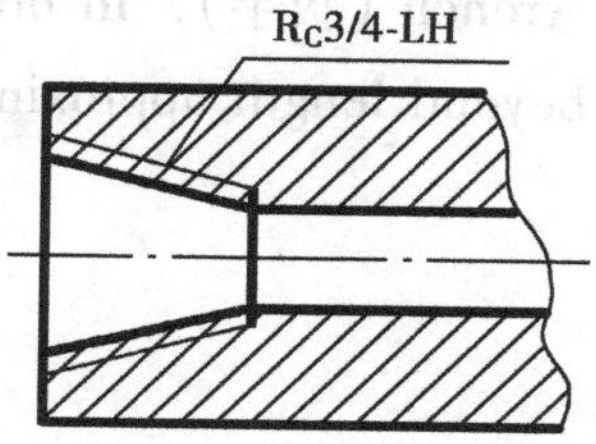 Rc3/4-LH
Taper pipe thread (Internal cylindrical thread)	R_p	1 ½	Left		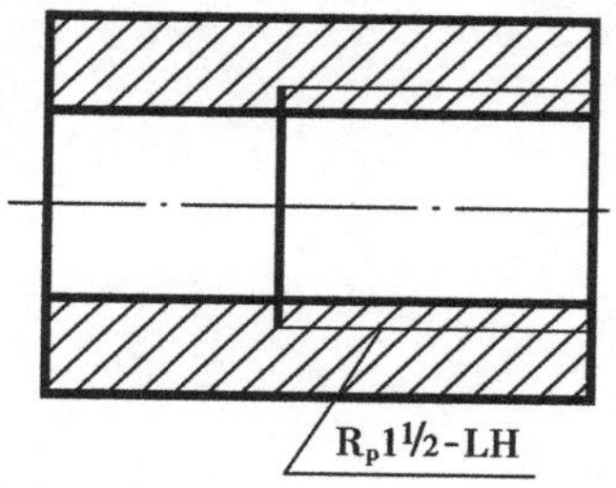 R_p1½-LH

8.2 Thread Fasteners and Stipulated Drawing（螺纹连接件及规定画法）

Thread fasteners（螺纹连接件）, such as bolt（螺栓）, stud（螺柱）, screw（螺钉）, nut（螺母）and washer（垫圈）are used widely in engineering. So, their forms and size are standardized by Chinese National Standard of Technical Drawing and they are made in standardized product factories（标准件工厂）. It is convenient for user to purchase them from standardized product shop by their notes. Manufacture catalogues provide information for drawing thread fasteners. Readers can also get the information from the Appendix 2 of this book. Thread joining（螺纹连接）is generally used for knockdown fastenings in engineering because there is always the necessity for assembling parts together, but they are expected to be disassembled easily.

8.2.1 Bolt joining（螺栓连接）

8.2.1.1 Composition

The thread fasteners used in a group of bolt joining include bolt, washer and nut. As shown in Fig. 8.20 (a), bolt has a hexagon head on one end and the other end is cut external thread partly. As shown in Fig. 8.20 (b), a nut is a threaded cylinder sleeve and it is used with a bolt to hold parts together. A washer is used to provide a bearing surface for smooth contact (Fig. 8.19 (c)).

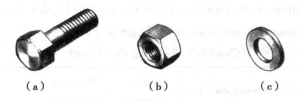

(a)　　　　　　(b)　　　　　(c)

Fig. 8.20　Thread parts used in bolt joining.

In joining as shown in Fig. 8.21, first two parts joined are drilled into clearance holes (光孔) whose diameter is 1.1 times the nominal size of the thread on the bolt. Second, the bolt is passed through clearance holes. Third, the washer is put. Lastly, the nut is screwed onto bolt with wrench (扳手). In order to ensure successful and reliable joining, the bolt is beyond nut. The beyond length approximates 0.3d, where d is the nominal size of the bolt.

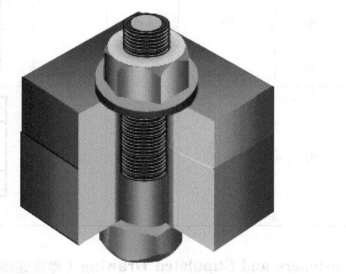

Fig. 8.21　Pictorial of bolt joining.

8.2.1.2　Stipulated drawing of bolt joining

Given: Height of upper plate and lower plate is δ_1 and δ_2 respectively, where $\delta_1 = 17$, $\delta_2 = 20$. Thread fasteners are as follows (Fig. 8.22):

Bolt　GB/T 5780—2000—M10 × L

Nut　GB/T 41—2000—M10

Washer　GB/T 95—1985—10—100HV.

Draw three views of bolt joining.

Solution is as follows.

Step1. Check Tables in Appendix 2.

　　Table 12: Bolt GB/T 5780—2000—M10 × L: $e = 17.59$, $K = 6.4$

Where letter e is the diameter of bolt head and letter K is the thickness of the bolt head.

　　Table14: Nut GB/T 41—2000—M10: $e = 17.59$, $m = 9.5$

Where letter e is the diameter of nut and letter m is the thickness of the nut.

　　Table17: Washer GB/T 95—1985—10—100HV: $d_2 = 20$, $h = 2$

Where letter d_2 is the diameter of washer and letter h is the thickness of the washer.

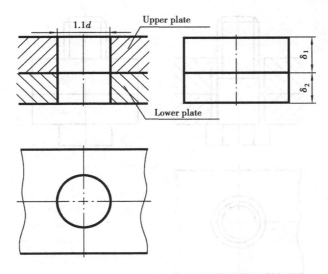

Fig. 8.22 Parts used to bolt joining.

Step2. Estimate the length of the bolt (估算螺栓的长度).

The estimated length of the bolt is summation of the height of upper plate and lower plate, the thickness of washer, the thickness of nut, as well as the 0.3d. Calculation formula is as follow.

$$L' = \delta_1 + \delta_2 + h + m + 0.3d$$
$$L' = 17 + 20 + 2 + 9.5 + 0.3 \times 10 = 51.5$$

Step3. Determine the nominal length of the bolt, L.

The nominal length of a bolt is the length except head. The length standardized by Chinese National Standard of Technical Drawing is called nominal length of the bolt.

Check length series of Table 12, select the nominal length of bolt, $L = 50$.

So, the note of the bolt is "Bolt GB/T 5780—2000—M10 × 50". Check Table 12 in Appendix 2 according to the $L = 50$ obtain the length of the thread on the bolt, $b = 26$.

Step4. Draw three views of bolt joining as shown in Figs. 8.23 (a) to (e).

Note 1. Although the curves produced on the bolt heads and nuts are hyperbolas (双曲线), these curves are always represented approximately by circular arcs (Fig. 8.23(f)) or straight

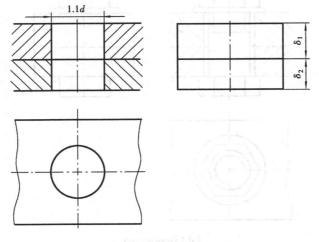

(a) Drawing two plates.

Fundamentals of Engineering Drawing

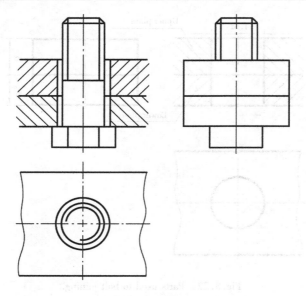

(b) Drawing bolt.

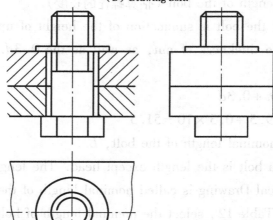

(c) Drawing washer.

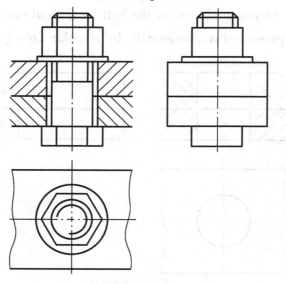

(d) Drawing nut.

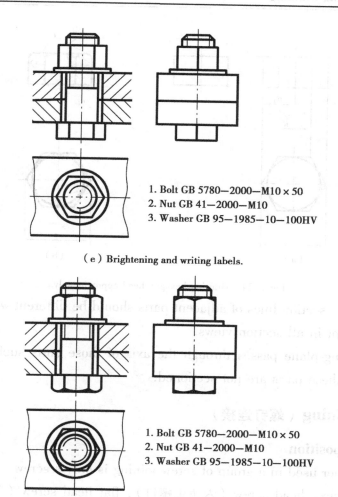

(e) Brightening and writing labels.

1. Bolt GB 5780—2000—M10×50
2. Nut GB 41—2000—M10
3. Washer GB 95—1985—10—100HV

(f) Drawing the head of the bolt and nut approximately.

1. Bolt GB 5780—2000—M10×50
2. Nut GB 41—2000—M10
3. Washer GB 95—1985—10—100HV

Fig. 8.23 Three views of bolt joining.

lines simply (Fig. 8.23(e)). Fig. 8.24 illustrates how to draw circular arcs. First, the bolt head is thought as a hexagonal prism simply, using the data, e and m, draw three views of the hexagonal prism (Fig. 8.24(a)). Secondly, in the front view, with the radius R (measured from the front view) draw large arc tangent to the top surface and the arc's center is located on the symmetric centerline. Thirdly, draw two small arcs with r (r and the center are determined by drawing as shown in the front view of Fig. 8.24 (b)). Fourthly, in the left view, with radius R_1 (measured from the front view) draw two arcs tangent to the top surface and the two arcs' centers are located on the middle line of the frontal frame and rear frame, respectively. Lastly, on the front view, draw chamfer at angle 30° with horizontal and draw chamfer circle on the top view, which is tangent to the hexagon. The approximate drawing of a nut is the same with a bolt head.

Note 2. Some regulations in joining drawing of thread fasteners.

1) The contacting surface of two parts is drawn by one thick line, but for noncontacting surfaces, they are drawn by two thick lines no matter whatever small the gap is. For example, the front view shown in Fig. 8.23 (e), the contacting surface of upper and lower plates is a thick line while there are two thick lines between clearance hole and rod of the bolt.

Fundamentals of Engineering Drawing

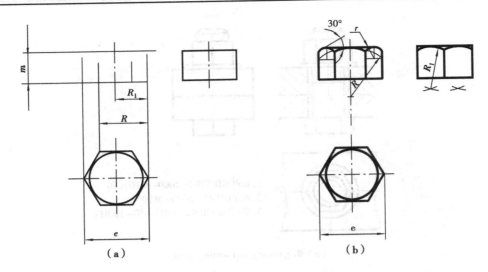

Fig. 8.24 Drawing hexagon-head approximately.

2) In the section, section lines of adjacent parts should be different while the section lines of one part should be kept in all section views.

3) When a cutting-plane passes through the axis of those parts such as bolt, nut, screw, stud or washer etc., these parts are not sectioned.

8.2.2 Screw Joining (螺钉连接)

8.2.2.1 Composition

The thread fastener used in a group of screw joining is only a screw. As shown in Fig. 8.25, a cap screw may be hex. head screw (六角头螺钉), flat head screw (平头螺钉), round head screw (圆头螺钉), fillister head screw (开槽螺钉) or hex. socket screw (内六角头沉孔螺钉).

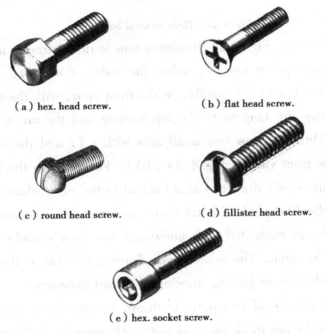

(a) hex. head screw.
(b) flat head screw.
(c) round head screw.
(d) fillister head screw.
(e) hex. socket screw.

Fig. 8.25 Cap screws.

As shown in Fig. 8.26, a cap screw is used to hold two parts together without a nut. The cap screw is passed through a clearance hole in one part and is screwed into a threaded hole in the other part with screwdriver (螺丝起子).

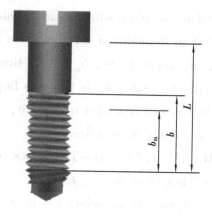

L: Nominal length
b: Threaded length
b_m: Engagement length

Fig. 8.26 Cap screw joining.

8.2.2.2 Stipulated drawing of screw joining

Given: The height of upper plate is δ_1, $\delta_1 = 20$, and the height of lower plate δ_2 is infinite. The material of lower plate is cast iron. Thread fastener is Screw GB/T 65—2000—M10 × L (Fig. 8.27).

Draw two views of screw joining.

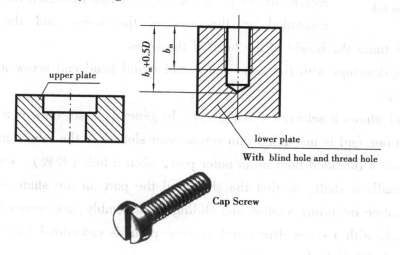

Fig. 8.27 Parts in a group of screw joining.

Solution is as follows.

Step1. Check Table 9 in Appendix 2.

Screw GB/T 65—2000—M10 × L: fillister head screw $d_k = 16$, $k = 6$, Where d_k is diameter of head of the screw while letter k thickness of head the screw.

Step2. Check Table 8 in Appendix 2 according to the $M = 10$: $d_2 = 18$, $t = 7$, as shown in Fig. 8.28, where d_2 is the diameter of the socket hole while letter t is the depth of the socket.

Step3. Estimate the length of the screw (估算螺钉的长度).

$$L' = \delta_1 - t + b_m$$

Where b_m is the engaged length of the screw (Fig. 8.26). It depends on the major diameter of thread on screw and the material of the plate with threaded hole. When a screw is screwed into steel plate, b_m is equal to the nominal diameter of the thread on screw, ie $b_m = d$. When a screw is screwed into cast iron, brass, or bronze plate, b_m is 1.5 times d, ie $b_m = 1.5d$; when it is screwed into aluminum, zinc, or plastic plate, b_m is twice as large as d, ie $b_m = 2d$.

In this problem, the screw is screwed into cast iron plant, so $b_m = 1.5d$.

The estimated the length of the screw is as follows:

$$L' = 20 - 7 + 1.5d = 20 - 7 + 1.5 \times 10 = 28$$

Step4. Determine the nominal length of screw, L.

In length series of Table 9 in Appendix 2, select the nominal length of screw, $L = 30$. So, the note of screw is "Screw GB/T 65—2000—M10 × 30". According to $L = 30$, the threaded length on the screw may be obtained from the note ② in Table 9. Namely, $b = L = 30$, this means that the rod of the screw is cut thread wholly.

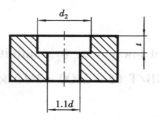

Fig. 8.28 Socket hole.

Step5. Draw two views of fillister screw joining as shown in Figs. 8.29 (a) to (d).

Besides, screwdriver slot (螺丝起子槽) is drawn as follows. In the circular view of the head, screwdriver slot is drawn at 45° with double thick line, without regard to true projection. In the rectangular view, a double thick line indicated the screwdriver slot is coincided on the axis of the screw and the length of it is approximately 2/3 times the height of the head of the screw.

Screw joining drawings with flat head screw and round head cap screw are shown in Figs. 8.30 (a) and (b).

Fig. 8.31 (a) shows a setscrew (紧定螺钉). In general, a setscrew is a headless threaded steel rod of which one end is machined into screwdriver slot while the other end is cone point. A setscrew screws into a threaded hole in an outer part, often a hub (轮毂), with its point against an inner part, usually a shaft, so that the shaft and the part on the shaft are held together to prevent relative motion including rotation and sliding. In assembly, a setscrew is screwed into hub with a threaded hole with a screw driver and its cone point is embedded (嵌入) shaft with a pit hole (凹坑). Fig. 8.30 (b) shows joining.

Chapter 8 Threads, Fasteners and Gears

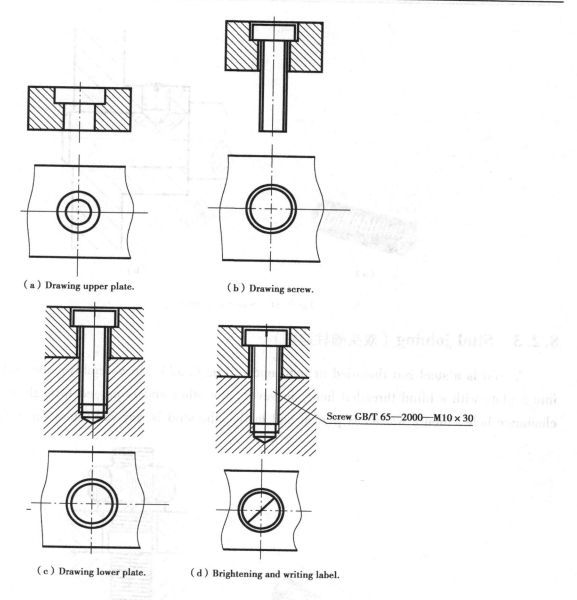

(a) Drawing upper plate. (b) Drawing screw.

(c) Drawing lower plate. (d) Brightening and writing label.

Fig. 8.29 Two view of fillister screw joining.

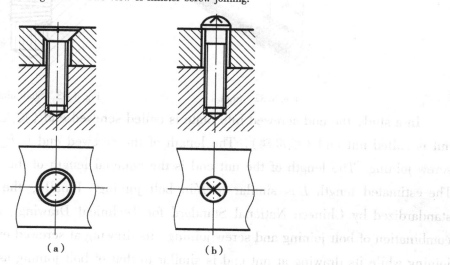

(a) (b)

Fig. 8.30 Screw joining drawings.

189

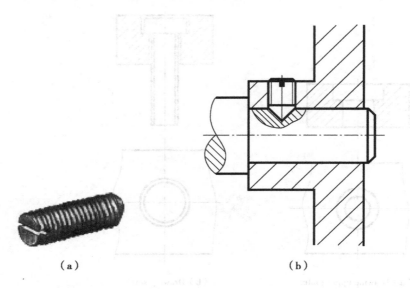

Fig. 8.31 Setscrew joining.

8.2.3 Stud joining (双头螺柱连接)

A stud is a steel rod threaded at both ends (Fig. 8.32). In assembly, one end is screwed into a plate with a blind threaded hole, however, the other end is passed through a plate with a clearance hole, then a washer is placed and lastly, the stud is screwed into a nut (Fig. 8.33).

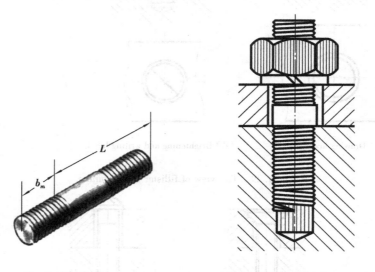

Fig. 8.32 Stud.   Fig. 8.33 Stud joining.

In a stud, the end screwed into plate is called screwed end (旋入端), but the end screwed a nut is called nut end (螺母端). The length of the screwed end is b_m, which is similar to b_m in screw joining. The length of the nut end is the nominal length of the stud, which is stood for L. The estimated length L' is similar to L' in bolt joining. Finally, the nominal length of stud is standardized by Chinese National Standard for Technical Drawing. Actually, stud joining is a combination of bolt joining and screw joining. Its drawing at screwed end is similar to that of screw joining while its drawing at nut end is similar to that of bolt joining as shown in Fig. 8.34.

Chapter 8 Threads, Fasteners and Gears

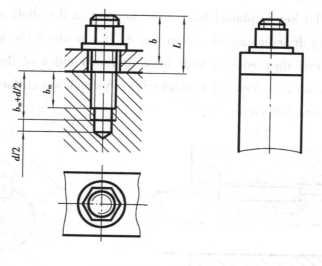

Fig. 8.34 Stud joining.

8.3 Keys and Key Joining (键和键连接)

Keys are used to prevent relative motion between shafts and wheels.

A key is a piece of metal and part of it is sunk into a groove, called the "key seat," cut in a shaft. The key then extends somewhat above the shaft and fits into a "keyway" cut in a hub (轮毂). After assembly, the key is partly in the shaft and partly in the hub, locking the shaft and wheel together so that one cannot rotate without the other (Fig. 8.35).

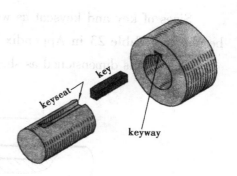

Fig. 8.35 Key nomenclature (术语).

8.3.1 Key types

Keys include flat keys (平键), woodruff keys (半圆键) and gib-head keys (钩头楔键).

Flat keys also divided into three types, type A (round ends key), type B (semi-round end key) and type C (square key) as shown in Figs. 8.36 (a) to (b). Letters L, b and h stand for key's width, depth and height, respectively. Flat keys have been standardized by Chinese National Standard for Technical Drawing. Standard sizes are given Tables 23 and 24 in Appendix 2.

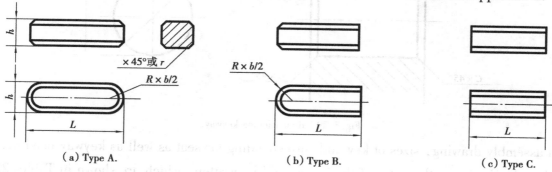

(a) Type A.　　　　　(b) Type B.　　　　　(c) Type C.

Fig. 8.36 Flat keys.

191

In key joining, a flat key is placed half in the keyseat in the shaft and half in the keyway in the hub. Flat key joining drawings are shown in Fig. 8.37, in which the key's front and rear sides are working surfaces, thus they contact with front and rear sides of the key seat and keyway. However, the key's top surface does not contact with keyway's top surface, so they appear as two thick lines in the front and left views.

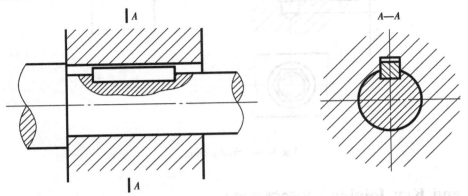

Fig. 8.37 Flat key joining.

Sizes of key and keyseat as well as keyway depend on the diameter of the shaft and they can be checked Table 23 in Appendix 2.

Key seat is dimensioned as shown in Fig. 8.38 and keyway is dimensioned as shown in Fig. 8.39.

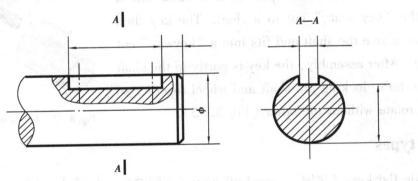

Fig. 8.38 Dimensioning keyseat.

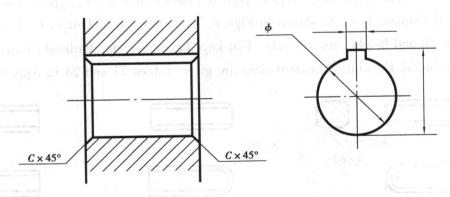

Fig. 8.39 dimensioning keyway.

In assembly drawing, sizes of key and corresponding keyseat as well as keyway need not to be dimensioned. However, the notes of the key should be written, which are shown in Tables 23 and 24 in Appendix 2.

Woodruff key is a flat segmental disk with round bottom (具有圆底的部分的圆盘状物体) as shown in Fig. 8.40. The keyseat is semicylindrical bottom and cut to a depth so that partial key can be put. Another partial key extends above the shaft in order to be driven into the hub. Woodruff key joining drawings are shown in Fig. 8.41. The working principle of the woodruff key is similar to that of the flat key.

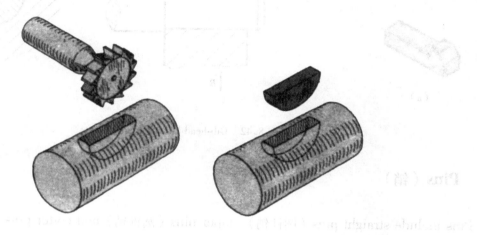

Fig. 8.40 Woodruff key, cutter, and keyseat.

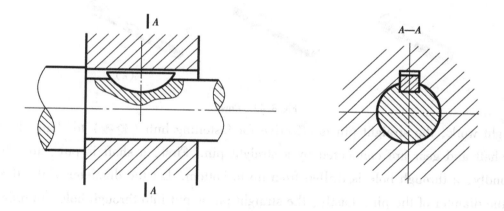

Fig. 8.41 Woodruff key joining.

The gib-head key (Fig. 8.42 (a)) is tapered on its upper surface and driven into hub to form a very secure fastening (安全,可靠的连接). The gib-head key joining drawings are shown in Fig. 8.42 (b), in which the key's the top surface is the working surface and it contacts with the top surface of the keyway, so the contacting surface appears as one thick line in the front and left views. However, the key's front and rear surfaces are nonworking surfaces, thus they do not contact with the front and rear surfaces of the keyseat and keyway, so noncontacting surfaces appear as two thick lines at front side and rear side in the left view.

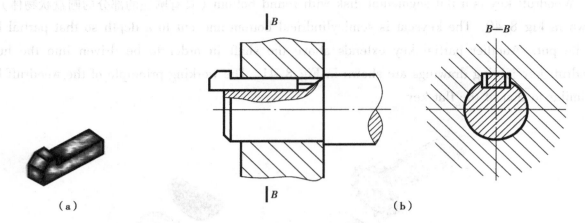

Fig. 8.42 Gib-head key and joining.

8.4 Pins (销)

Pins include straight pins (圆柱销), taper pins (圆锥销) and cotter pins (开口销) as shown in Figs. 8.43 (a) to (c).

Fig. 8.43 Pins.

For light work, the straight pin is effective for fastening hub (轮毂) to shaft. In Fig. 8.44 (a), the shaft and gear are connected by a straight pin. First, a shaft is put into a hub of the gear. Secondly, a through hole is drilled from up to bottom. Let the diameter of the through hole equale to the diamter of the pin. Lastly, the straight pin is put into through hole. In order to avoid that the straight pin intervenes (干涉) other parts when the gear rotating, it is necessary to hide pin into a hub of the gear.

Taper pins are conical in shape and used for a variety of purpose, chief of which is to keep two parts in a fixed position or to pressure alignment (对齐) between parts. In order to disassemble (拆卸), taper pin is higher than adjacent objects. As shown in Fig. 8.44 (b), when the pin is beaten from bottom to top with a hammer, the taper pin is disassembled easily from the adjacent objects.

Cotter pins has two legs in shape as for a clip (夹子), and are used to prevent nuts from loose. Generally, a screw thread holds securely unless the parts are subject to (易于…) impact and vibration (碰撞,振动), as in a railroad or an automobile engine. When a slotted nut (开槽螺母) shown in Fig. 8.44 (c) is screwed onto a bolt with a hole (带孔螺栓) and then a cotter is traversed the rear slot at the slotted nut, the hole of the bolt as well as the front slotted at slotted nut, two legs are divided to prevente from loosening as shown in Fig. 8.44 (d).

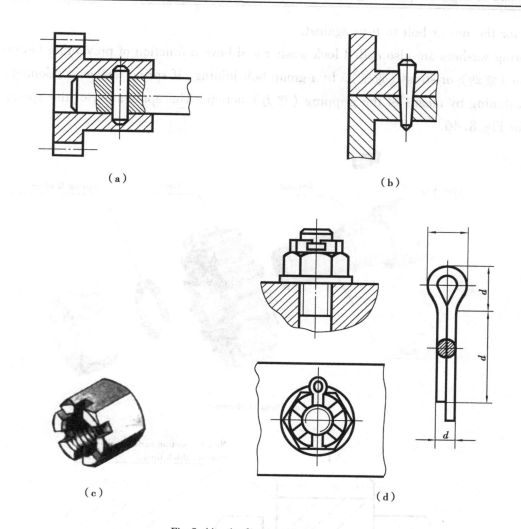

Fig. 8.44 Application of pins.

Size of each pin and its note are shown in tables in Appendix 2.

8.5 Washers (垫片)

In engineering, there are two kinds of washers, plain washers (平垫片) and spring washers (弹簧垫片) as shown in Fig. 8.45.

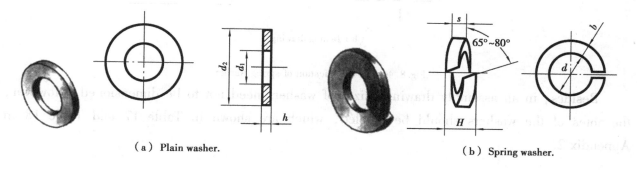

(a) Plain washer.　　　　　　　　(b) Spring washer.

Fig. 8.45 Washers.

Plain washers are commonly used in the assembly of nuts and bolts to provide a smooth

195

surface for the nut or bolt to turn against.

Spring washers are also called lock washer and have a function of preventing loosening due to vibration (震动) or stress (重压). In a group bolt joining, if spring washer is adopted, it make a quick fastening by means of the gripping (张力) action. The application of the spring washer is shown in Fig. 8.46.

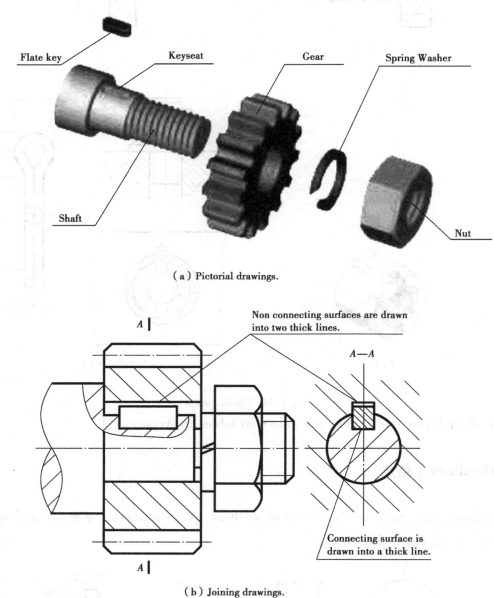

(a) Pictorial drawings.

(b) Joining drawings.

Fig. 8.46 The application of spring washer.

Besides, in an assembly drawing, sizes of washers need not to be dimensioned. However, the notes of the washers should be labeled, which are shown in Table 17 and Table 18 in Appendix 2.

8.6 Gears (齿轮)

8.6.1 Introduction

Gears are toothed wheels. Two gears mesh (啮合) mutually to transmit torque (扭矩), rotary motion and power from one shaft to another.

There are numerous variations, but the basic forms are spur gears (直齿圆柱齿轮), for transmitting torque between parallel shafts; spur gear and rack (齿轮齿条), for changing rotary motion to linear motion; bevel gears (锥齿轮), for shafts whose axes intersect; and worm gears (蜗轮蜗杆), for nonintersecting shafts at right angles to each other as shown in Figs. 8.47 (a) to (d). Of all types, the spur gear is the simplest. The spur gear is taken under discussion.

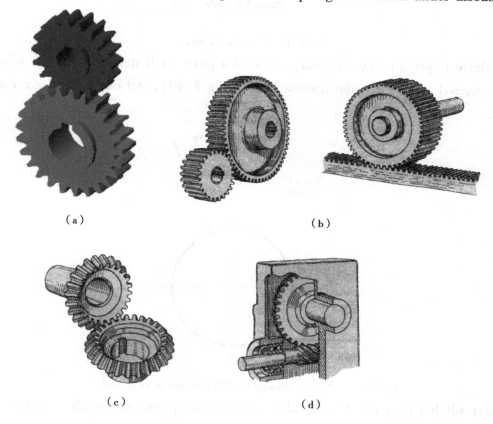

Fig. 8.47 Basic gear forms.

8.6.2 Terms

The follows terms and formulae (formula 的复数) are used to describe the parts of the spur gear, several of which are shown in Fig. 8.48.

Gear teeth (轮齿) The teeth of a gear are designed to fit the tooth space of the meshing gear. Capital letter Z standards for the number of teeth on a gear. The most common form of the tooth flank (齿廓) is the involute (渐开线), and when it is made in this form, the gear is known as involute gear (渐开线齿轮).

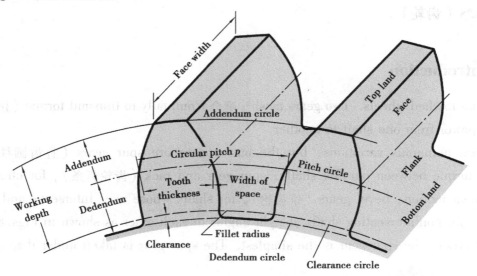

Fig. 8.48 Terms of the spur.

Pitch circle (分度圆) The imaginary circle of a gear, as if it were a friction wheel without teeth that contacted another circular friction wheel (Fig. 8.49). All calculations are based on the pitch circle.

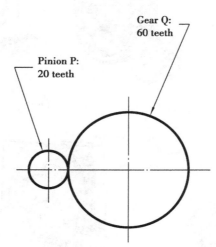

Fig. 8.49 Schematic drawing (示意图) of spur gears in mesh.

Circular pitch (齿距 P) The circular arc from one point on a tooth to the corresponding point on the next tooth measured along the pitch circle.

Circular thickness (齿厚 S) The circular arc across a tooth measured along the pitch circle.

Pitch diameter (分度圆直径 d) The diameter of the pitch circle.

The perimeter (周长) of the pitch circle is calculated as follows:

$Z \times P = \pi \times d$

So, the diameter of the pitch circle is calculated as follows:

$d = P/\pi \times Z$

Where P/π is defined as module (模数), which is denoted by letter m. The unit of the module is millimeters.

Thus the diameter of the pitch circle is also calculated as follows:

$d = m \times Z$

Module m is a very important parameter (参数) for designing and machining a gear. The size of teeth is specified by the module. For two gears to mesh, they must have the same module, that is, gears to be run together must be cut with the same module' machine tool. In order to design and machine conveniently, modules have standardized as shown in Chart 8.6. Note that series 1 is selected preferentially (优先) and the series located in bracket do not be selected as much as possible.

Chart 8.6　Modules（GB 1357—1987）

Series 1	1	1.25	1.5	2	2.5	3	4	5	6	8	10	12	16	20	25	32	40	50
Series 2	1.75	2.25	2.75	(3.25)	3.5	(3.75)	4.5	5.5	(6.5)	7	9	(11)	14	18	22	28	36	45

The **addendum** (齿顶高 h_a) is radial distance between the top land and pitch circle.

The **dedendum** (齿根高 h_f) represents the radial distance from the bottom land to the pitch circle.

The **whole depth** (齿全高 h) is the sum of addendum and dedendum.

The **clearance circle** (间隙圆) represents a circle in teeth space tangent to the addendum circle of the mating gear (在齿槽中,与啮合齿轮的齿顶相切的圆).

The **clearance** (间隙 f) represents the amount by which the dedendum in a given gear exceed (超越) the addendum of the mating gear.

Note that clearance is required to prevent the end of the tooth of one gear from riding (骑) on the bottom of the mating gear.

The **working depth** (工作高度) is the difference between the whole depth and clearance.

The **addendum diameter** (齿顶圆直径 d_a) is the maximum diameter of a gear across its teeth.

The **dedendum diameter** (齿根圆直径 d_f) is the diameter of a gear measured from the bottom of its gear teeth.

As shown in Fig. 8.50, when two gears are in mesh, two pitch circles roll on one another without slipping at the fixed point called pitch point P (节点). Note that the pitch point P always passes though the line of centers $O_1 O_2$. Line ab is the common tangent (公切线) through the pitch point and line cd is the common normal (公法线) to the teeth that are in contact at pitch point.

The **pressure angle** (压力角) is an angle between the common normal and common tangent at pitch point, which is standardized by 20°.

The **gear ratio** is the ratio of revolutions and it is the inverse of the number of the teeth.

$i = n_1/n_2$ or $i = n_1/n_2 = Z_2/Z_1$.

Where

n_1 = revolutions per unit of time for driver gear (主动轮)

n_2 = revolutions per unit of time for driven gear (从动轮)

Z_1 = numbers of teeth for driver gear

Z_2 = numbers of teeth for driven gear

Subscripts (下标) 1 and 2 refer to driver and driven gears, respectively.

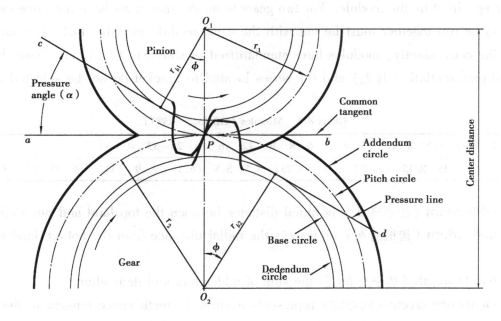

Fig. 8.50 Meshing gears.

The **center distance** (中心距 a) is the distance between the centers of the two gears in mesh. For a pair of meshing spur gears, when they are installed correctly (no errors), the center distance is calculated as follows:

$$a = d_1/2 + d_2/2 = m \cdot Z_1/2 + m \cdot Z_2/2 = m/2 \cdot (Z_1 + Z_2)$$

The number of teeth and module are two basic parameters. The other parameters of the standard spur gear can be determined by the two basic parameters as shown in Chart 8.7.

Chart 8.7 Parameters for standard spur gear

Basic parameters: module m, number of teeth Z		
Name	Code	Formulae
Addendum (齿顶高)	h_a	$h_a = m$
Dedendum (齿根高)	h_f	$h_f = 1.25\ m$
Whole depth (齿全高)	h	$H = 2.25\ m$
Pitch diameter (分度圆直径)	d	$D = mZ$
Addendum diameter (齿顶圆直径)	d_a	$d_a = m(Z + 2)$
Dedendum diameter (齿根圆直径)	d_f	$d_f = m(Z - 2.5)$

8.6.3 Drawing spur gears

8.6.3.1 Drawing a single spur gear

As shown in Fig. 8.51 (a), let the gear's axis be perpendicular to plane W. Fig. 8.51 (b) shows a conventional drawing of a spur gear. The gear appears as rectangles in the front view in which the gear is sectioned, but gear teeth are not sectioned. The gear appears as circles in the left view. Addendum circle (齿顶圆) and addendum line (齿顶线) are drawn with thick line. Pitch circle (分度圆) and pitch line are (分度线) drawn with center line. In the left view, dedendum circle (齿根圆) is omitted, however, in the front view, dedendum line (齿根线) is drawn with thick line.

(a) Pictorial drawing of a spur gear.

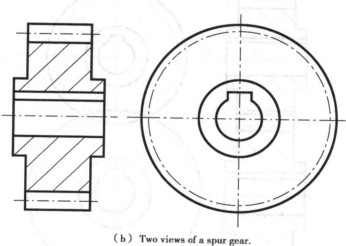

(b) Two views of a spur gear.

Fig. 8.51 Conventional drawing of a spur gear.

8.6.3.2 Drawing two spur gears in mesh

Fig. 8.52 (a) shows the pictorial drawing of two spur gears in mesh and their conventional drawing is shown in Fig. 8.52 (b). Except mating area, other drawings are similar to a single gear. The mating area appears as five lines in the front view because of pitch line in common. Note that when one gear's addendum line is drawn with thick line, the other one is drawn with hidden line. The mating area appears as two tangent pitch circles in the left view. Again, addendum circles may be drawn wholly or omitted in the mating area. Lastly, dedendum circles are omitted.

Because dedendum is higher than addendum by 0.25 m, the clearance between the dedendum line of one gears and the addendum of the other gear should be represented with two thick lines as shown in Fig. 8.52 (c).

(a) Pictorial drawing of two spur gears in mesh.

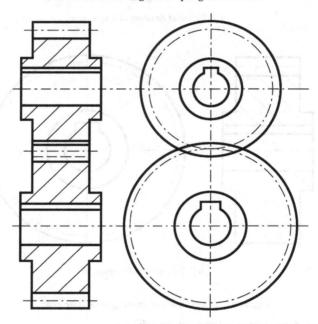

(b) Two views of two spur gears in mesh.

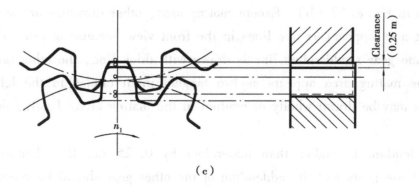

(c)

Fig. 8.52 Conventional drawing of two spur gears in mesh.

Chapter 9　Detail Drawings（零件图）

Any machine is assembled with many parts（零件）. A part is a basic unit to make a machine. A detail drawing is the drawing of a single part, giving a complete and exact description of its shape, dimensions and the desired technical requirements.

This chapter includes contents of a detail drawing, selecting views for a detail drawing as well as technical requirements in machining. Besides, typical part's detail drawings will be introduced. Lastly, the method for reading a detail drawing is illustrated.

9.1　Contents

Detail drawings are important technical documents that embody the designers' ideas. A successful detail drawing will tell the workers simply and directly the shape, size, material; what shop operations are necessary; what limits of accuracy must be observed. Based on it, manufacturing and inspecting a part can be carried out.

Fig. 9.1 is a detail drawing of transmission shaft（传动轴）. Let's take it as an example to illustrate what contents a detail drawing should include.

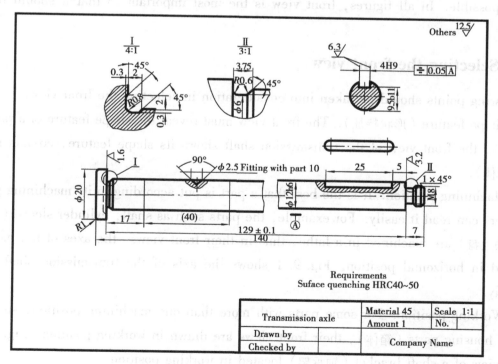

Fig. 9.1　Detail drawing of transmission shaft.

Figures　Some figures such as views, sections, cross-sections and zoom views are used to reveal the outside and inside structures of a part clearly and precisely.

Dimensions   All necessary dimensions for manufacturing and inspecting a part, including dimension datums (基准) in width, depth and height, location dimensions and size dimensions as well as general dimensions and so on.

Technical requirements   Technical requirements in manufacturing includes surface roughness, tolerance and heat-treatment and so on. Among them, surface roughness is more important since finished surfaces should be pointed out and all necessary shop (车间) operations shown in detail drawing.

Title block (标题栏)   A table is used to fill in the name of the part, the name of the company, the material of the part, scale for drawing and date for drawing as well as the signatures of the designer, drafter and checker.

9.2   Selecting Views

If a part is expressed on a drawing paper, all representing methods such as primary views, removed views, partial views, auxiliary views, full sections, half sections, broken-out sections, removed cross-sections, revolved cross-sections and zoom views as well as conventional practices (惯例表达) are all applicable if necessary. The principle to determine the best representing scheme is that the internal and external structures of a part are represented clearly and figures are as few as possible. In all figures, front view is the most important so that it should be built in advance.

9.2.1   Selecting the front view

Following points should be taken into consideration in selecting the front view.

1) Shape feature (形状特征). The front view must reveal the shape feature of a part mostly. In Fig. 9.1, the front view of the transmission shaft shows its shape feature, co-axis revolutions (同轴回转体).

2) Machining position. It is the best that a part is put according to its machining position so that workers can read it easily. For example, the parts such as shaft, cylinder sleeve (套筒) and flange (法兰盘) are machined in a lathe, thus in their front views, the axes of the parts usually are located in horizontal position. Fig. 9.1 shows the axis of the transmission shaft is located horizontally.

3) Working position. For some parts with more than one machining position, such as forks (叉) and housing parts (箱体), their front views are drawn in working position. Fig. 9.2 shows the front view of a shaft bracket (轴承架) located in working position.

Above three aspects should be considered comprehensively. If one part has several machining positions, the front view is drawn generally based on its shape feature. Besides, try to make more surfaces in the part reveal true sizes in the front view.

9.2.2 Selecting other views

After drawing the front view, other views are considered based on showing the part completely and clearly. According to the demand for representing, other views may be primary views, removed views, partial views, auxiliary views, all kinds of sections and cross-sections and so on.

To represent a part clearly, it is advisable to consider many expressing schemes and try to find the best one in them. Fig. 9.2 lists two expressing schemes of a bracket. Scheme (b) is better than (a) for it is more clear and simple. As shown in Fig. 9.2 (c) is pictorial drawing of the bracket.

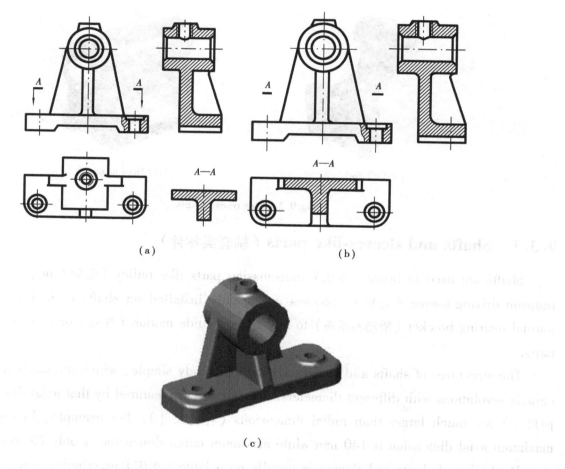

Fig. 9.2 Selecting a drawing scheme for a shaft bracket.

9.3 Typical Parts

In engineering practice, parts can be classified into four types: shafts and sleeves (套), flanges (法兰盘) and covers (端盖), forks and brackets (叉架) and housing parts (箱体) as shown in Figs. 9.3 (a) to (d).

(a) Shafts and sleeves. (b) Flanges and covers.

(c) Fork. (d) Housing part.

Fig. 9.3 Four types of parts.

9.3.1 Shafts and sleeves-like parts (轴套类零件)

Shafts are used to brace (支承) transmission parts like pulley (带轮) or gear (齿轮) and transmit driving torque (力矩). Sleeves are usually installed on shafts or in the holes of the journal bearing bracket (滑动轴承座) to fix, brace, guide motion (导向) or protect transmission parts.

The structures of shafts and sleeves are comparatively simple, which are made up of various co-axis revolutions with different diameters. These parts are featured by that axial dimensions (轴向尺寸) are much larger than radial dimensions (径向尺寸). For example, in Fig. 9.1, the maximum axial dimension is 140 mm while maximum radial dimension is only 20 mm.

Machining of shafts and sleeves is usually on a lathe (车床) or grinding machine (磨床), so, the front view is formed based on machining position where the part's axis is positioned horizontally and perpendicular to plane W. In general, in the front view, broken-out sections are adopted to exhibit interior structures such as pit holes (凹坑) and keyseats (轴上键槽). Besides, removed cross-sections are also chosen to reveal cross sections of keyseats. For example, in Fig. 9.1 there are five figures in the detail drawing including front view, partial view, removed cross-section, zoom view and zoom section. The main shape of the transmission shaft is represented by front view in which two broken-out sections and breaking drawing (断裂画法) are adopted to reveal the height of the pit hole and keyseat. The true shape and the depth of the keyseat are represented by a partial view adjacent the front view. Besides, a removed cross-section is used to dimension and label technical requirements about the key seat, such as 4H9, 95b11, etc. In order to represent tool escape (退刀槽) clearly, zoom view shown in figure II is adopted. Every figure having its representing purpose, analying of the rest figures is left to readers.

9.3.2 Flange-like parts (盘类零件)

Flange-like parts include gear, hand wheel and cover (端盖) etc. Their structures are also comparatively simple, which are made up of some co-axis revolutions and square plates (方板). Their blanks (毛坯) are from casting (铸造) or forging (锻造), but most of surfaces are machined. These parts are featured by that their axial dimensions are smaller than radial dimensions or side dimensions (端面尺寸). For example, Fig. 9.4 shows a detail drawing of an end cover (端盖) which is constructed by co-axis revolutions and a square plate. Its maximum axial dimension is 58 mm while maximum side dimension is 115 mm.

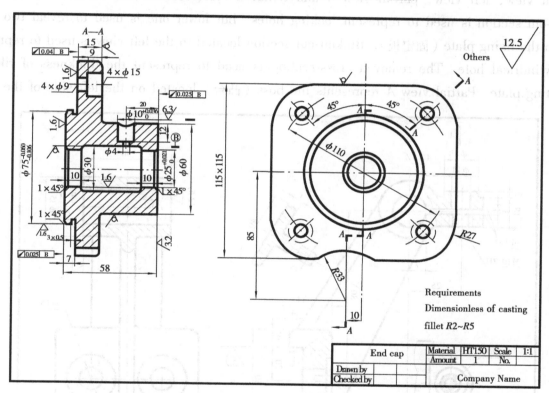

Fig. 9.4 Flange-like parts.

In machining, these parts' axes are located horizontally. So, the front view is formed based on machining position whose axis is horizontal and perpendicular to plane W. Usually, two views are selected to represent the parts. A full section frequently is adopted as front view to reveal interior structures mostly. A view is used as left view or right view to represent external structures. Fig. 9.5 is the pictorial drawing of the end cap shown in Fig. 9.4.

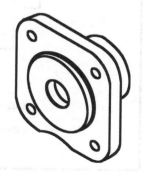

Fig. 9.5 Pictorial drawing of the end cap.

9.3.3 Fork and bracket-like parts (叉架类零件)

Fork and bracket-like parts include forks, connecting rods (连杆) and trestles (支架). Their blanks are from casting or forging and only a few surfaces need to be machined, which are surfaces in holes and some important side surfaces. Their structures are irregular and shapes are intricate (复杂). The manufacturing processes are diverse (变化的), so the front view is formed based on working position and shape feature. Usually the front view is drawn by broken-out section to reveal interior and exterior structures simultaneously. To represent their oblique structures, auxiliary views, auxiliary sections are applied. Revolved cross-sections and removed cross-sections are used to represent thickness of rib and connecting plate. For example, Fig. 9.6 is a detail drawing of a fork stand (叉架). There are four figures in the detail drawing, including the front view, left view, partial view A and removed cross-section. In the front view, upper broken-out section is used to represent locking holes, but lower one is used to reveal two sunk holes in the fixing plate (固定板). Broken-out section located in the left view is used to represent large cylindrical hole. The removed cross-section is used to represent the thickness of rib and connecting plate. Partial view A represents the boss (凸台) located on the left side of the large

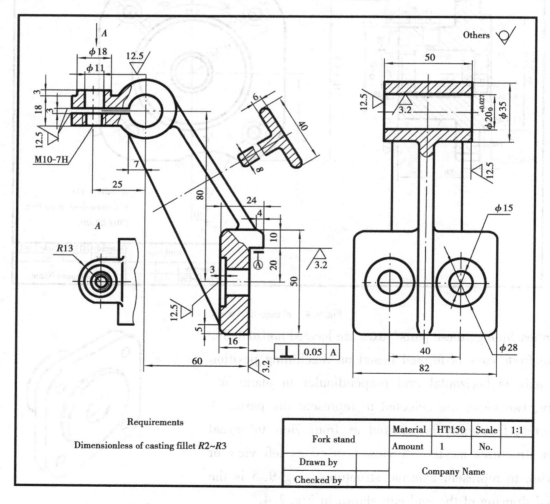

Fig. 9.6 Fork bracket-like parts.

cylindrical sleeve. After analyzing the front view, the left view, partial view A as well as the removed cross-section, we can figure out the object's internal and external shape of the fork stand. Fig. 9.7 is the pictorial drawing of the fork stand.

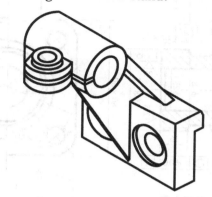

Fig. 9.7 Pictorial drawing of the fork stand.

9.3.4 Housing-like parts (箱体类零件)

Housing-like parts are used to contain and brace other parts, thus, part's middle is empty. Their blanks are from casting and only a few surfaces are machined. In general, there are bosses, pit holes, thread holes, sunk holes, ribs in them. So, their structures are quite complex.

Because of the diversity of machining positions, the front view is formed based on working position and shape feature. Besides, three or more primary views are needed to represent outside shape clearly. Proper sections are absolutely necessary to express inside structures. Other partial structures can be expressed by partial views and cross-sections and so forth.

Fig. 9.8 shows a detail drawing of a valve body (阀体). Three views are adopted. The front view is a full section in order to express its inner structure. The top view is used to express the outside shape. The left view is a half section because the valve body is approximately symmetrical in depth direction except a rectangular boss in front. The shape of the rectangular boss is represented by a partial view B. Fig. 9.9 is a pictorial drawing of the valve body.

9.4 Technical Requirements

9.4.1 Surface quality

9.4.1.1 Introduction

Simple dimensioning of the width, depth and height of a part indicates little of the surface condition of that part. A surface could be in an as-cast (铸造) condition or it could be machined by various machining methods. The surface being in an as-cast condition is called rude surface (毛面), however the surface machined by various machining method is called finished surface (加工面). A designer must be able to specify surface quality so that the part can be made as desired. Normally, the ideal finished surface is the roughest one that will do the job satisfactorily.

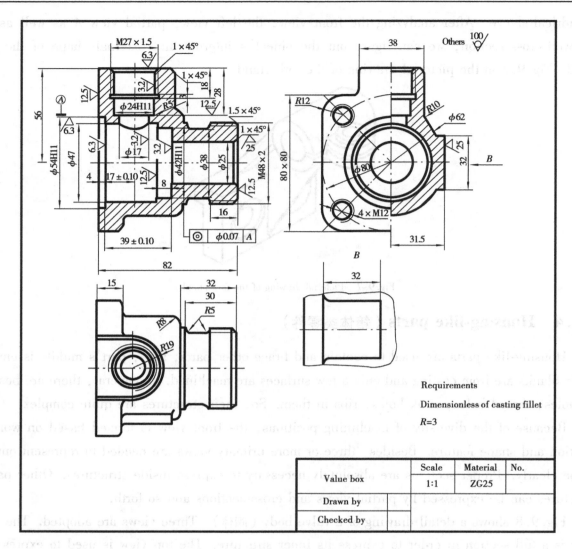

Fig. 9.8 Housing body like-parts.

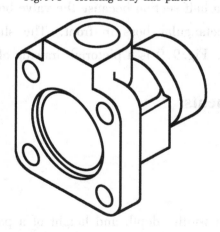

Fig. 9.9 Pictorial drawing of the value box.

9.4.1.2 Surface roughness

As shown in Fig. 9.10, any surface has minute (微小的) peaks and valleys in a microscope, the height of which is termed "surface roughness". Usually, we adopt surface roughness height parameters (表面粗糙度高度参数) to assess (评定) surface quality.

As shown in Fig. 9.11, surface roughness height parameters are the arithmetical average deviation（算术平均偏差）from the average height line（in micrometers μm）and it is represented by letter Ra.

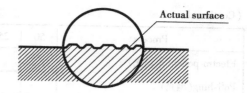

Fig. 9.10 Surface roughness.

$$Ra = \frac{1}{l}\int_0^l |y(x)|\,\mathrm{d}x$$

Where letter l means the sampling length（取样长度）while "$y(x)$" the deviation（偏距）.

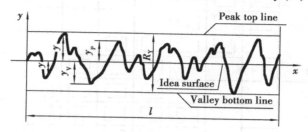

Fig. 9.11 Surface roughness height parameters.

Roughness heights actually produced by common tooling are shown in Chart 9.1.

Chart 9.1　Roughness average Ra 　　units: μm

Process	50	25	12.5	6.3	3.2	1.6	0.8	0.4	0.2	0.10	0.025	0.012
Flame cutting（火焰切割）	☐	☐	☐									
Snagging（粗磨）	☐	☐	☐	☐								
Sawing（锯切）		☐	☐	☐	☐							
Planing, shaping（平面刨光）		☐	☐	☐	☐	☐	☐					
Drilling（钻）			☐	☐	☐	☐						
Chemical milling（化学铣）				☐	☐	☐	☐					
Elect. discharge mach.（放电加工）				☐	☐	☐	☐					
Milling（磨）			☐	☐	☐	☐	☐	☐				
Broaching（扩孔）					☐	☐	☐	☐				
Reaming（铰孔）					☐	☐	☐	☐				
Electron beam（电子束）					☐	☐	☐	☐				
Laser（激光）					☐	☐	☐	☐				
Electro-chemical（电化学）					☐	☐	☐	☐	☐	☐		
Boring, turning（镗扩）		☐	☐	☐	☐	☐	☐	☐	☐	☐		
Barrel finishing（滚筒电镀法）						☐	☐	☐	☐			
Electrolytic grinding（电研磨）							☐	☐	☐			
Roller burnishing（滚轴磨光）							☐	☐				
Grinding（研磨）						☐	☐	☐	☐	☐		
Honing（珩磨）							☐	☐	☐	☐		

(Continued)

Process	50	25	12.5	6.3	3.2	1.6	0.8	0.4	0.2	0.10	0.025	0.012
Electro-polish(电抛光)						□	□	□	□	□	□	
Polishing(抛光)							□	□	□	□	□	
Lapping(精抛光)								□	□	□	□	□
Superfinishing(超级研磨)							□	□	□	□	□	
Sand casting(沙铸)	□	□	□									
Hot rolling(热轧)	□	□										
Forging(锻造)		□	□	□								
Perm mold casting(永久模铸造)			□	□	□							
Investment casting(耐火铸造)			□	□	□							
Extruding(挤压)			□	□	□							
Cold rolling, drawing(冷,轧)				□	□	□						
Die casting(硬模铸造)				□	□	□						

9.4.1.3 Surface roughness symbols

Surface roughness symbols are shown in Fig. 9.12, where $H_1 = 1.4h$, $H_2 = 3h$ while letter h is the height of the characters in the detail drawings.

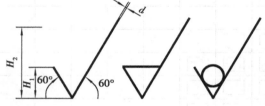

Fig. 9.12 Surface roughness symbols.

Symbol ∨ is a basic surface roughness symbol and it indicates that a finished surface may be produced by any method.

Symbol ∇ is a surface roughness symbol of material removal. The horizontal bar indicates that material removal by machining is required to produce the surface.

Symbol ∨̥ is a surface roughness symbol of material removal prohibited (禁止). The circle in the vee expresses that the surface must be produced by processes such as casting, forging, hot finishing, cold finishing, die casting, without subsequent (后来的) removal of material.

9.4.1.4 Surface roughness codes

GB/T 131—1993 stipulates that surface roughness code consists of surface roughness symbol and surface roughness height parameter. Some examples of surface roughness codes are shown in Chart 9.2.

Chart 9.2 Examples of surface roughness codes

Code	Meaning	Code	Meaning
3.2 ∨	Surface may be produced by any method and the upper limit of Ra is 3.2 μm.	3.2 ∇	Material is removed by machining and the upper limit of Ra is 3.2 μm.
3.2 ∨̥	Removal of material is prohibited and the upper limit of Ra is 3.2 μm.	3.2 / 1.6 ∇	Material is removed by machining. The upper limit of Ra is 3.2 μm while the lower limit of Ra is 1.6 μm.

Code	Meaning	Code	Meaning
3.2max ∀	Surface may be produced by any method and the maximum of Ra is 3.2 μm.	3.2max ∇	Material is removed by machining and the maximum of Ra is 3.2 μm.
3.2max ⌀	Removal of material is prohibited and the maximum of Ra is 3.2 μm.	3.2max 1.6min ∇	Material is removed by machining. The maximum of Ra is 3.2 μm while the minimum of Ra is 1.6 μm.

9.4.1.5 Labeling surface roughness codes

First, in drawing a surface roughness symbol, the long leg turns around the short leg in counterclockwise. Second, the head of number representing roughness average value should be written above horizontal bar. Thirdly, surface roughness codes are placed on the visible contour lines, dimension extensions or leader lines of the finished surfaces. Note that each surface roughness code should be pointed to the surface from empty to solid as shown in Fig. 9.13.

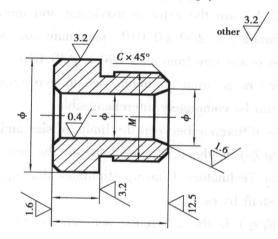

Fig. 9.13 Labeling surface roughness codes.

9.4.2 Tolerance and fit

In engineering practice, it is impossible to make anything to exact size. Fortunately, exact sizes are not needed. However the need is for varying degrees of accuracy according to functional requirement. The varying range of exact size is known as dimension tolerance or tolerance simply. Tolerance is actually a permissible error in size.

Interchangeable manufacturing (互换性生产) allows parts made in wide separated localities to be brought together for assembly. Without interchangeable manufacturing, modern industry could not exist, and without dimension tolerance, interchangeable manufacturing could not be achieved.

9.4.2.1 Tolerance
①**Terms**

Take the dimension $\phi 30 \pm 0.010$ as example to illustrate terms as follows (Fig. 9.14).

Basic size(基本尺寸) is the theoretical size from designing, for example $\phi 30$.

Actual size(实际尺寸) is the measured size of the finished part.

Fundamentals of Engineering Drawing

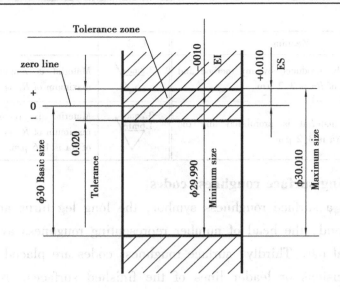

Fig. 9.14 Terms.

Limits of size (极限尺寸) are the extreme maximum and minimum sizes specified by a tolerance dimension. For example, in $\phi30 \pm 0.010$, maximum size is $\phi30.010$ while minimum size is $\phi29.990$. Actual size is any size from $\phi29.990$ to $\phi30.010$.

If products' actual sizes range from minimum size to maximum size, these products are eligible (合格的) and they can be completely interchangeable.

Deviation (偏差) is the difference between the limits of size and basic size.

Upper deviation (上偏差) is the difference between the maximum size and basic size. Chinese National Standard of Technology Drawing stipulates that upper deviation for a hole is indicated by ES while for a shaft by es.

Lower deviation (下偏差) is the difference between the minimum size and basic size. Chinese National Standard of Technology Drawing stipulates that lower deviation for a hole is indicated by EI while for a shaft by ei.

In Fig. 9.14(a), Upper deviation ES = 30.010 − 30 = +0.010
Lower deviation EI = 29.990 − 30 = −0.010

Fundamental deviation (基本偏差) is the deviation closest to the basic size.

If take one of the dimension extension lines of the basic size as zero line (零线), fundamental deviation may be superposed (重叠) on the zero line (zero deviation) or may be higher than zero line (positive deviation) or may be lower than zero line (negative deviation).

Tolerance (公差) is the total amount that a specific dimension is permitted to vary. It is also the difference between the maximum limit size and the minimum limit size of a part or the difference between the upper deviation and the lower deviation for the dimension. Chinese National Standard of Technology Drawing stipulates that tolerance is indicated by γ. For example, in Fig. 9.14 (a), $\gamma = 30.010 - 29.990 = 0.020$ or $\gamma = +0.010 - (-0.010) = 0.020$.

A dimension with tolerance consists of basic dimension followed by upper deviation and lower deviation. For example, $\phi30 \pm 0.010$ or $\phi30_{0}^{+0.002}$.

Fundamental deviation series (基本偏差系列) consist of 28 fundamental deviations.

Chinese National Standard of Technology Drawing stipulates that the fundamental deviation of a hole is represented by capital Latin letters while small Latin letters (小写字母) is used to represent the deviation of a shaft. As shown in Fig. 9.13, from shaft a, shaft b... to shaft zc, the shafts total 28; from hole A, hole B ... to hole ZC, the holes total 28.

International tolerance grade (IT 标准公差等级) is a set of tolerances that varies with the basic size and provides a uniform level of accuracy within a given grade. In all, there are 20 IT grades—IT01, IT0 and IT1 through IT18. The higher the numeral is, the larger the tolerance is, and the larger varying range of the size is. Practical use of the IT grades is as follows:

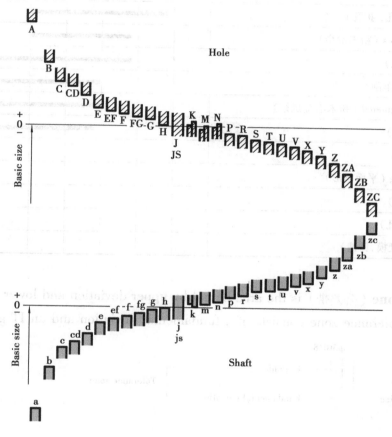

Fig. 9.15 A series of deviations.

IT01—IT1—used for measuring tools.

IT2—IT5—used for precision parts.

IT6—IT12—used for fit parts.

IT6—IT18—used for important parts.

IT grades reflect the degrees of accuracy of a dimension. Manufacturing costs increase as tolerances become smaller. A tolerance should be as large as possible without interfering (妨碍) with the function of the part to minimize production costs.

Chinese National Standard of Technology Drawing stipulates each IT grade listed in Table 27 in Appendix 2.

In engineering practice, different IT grade should be corresponded different machining processes as shown in Chart 9.3.

Chart 9.3 **IT grades related to machining processes.**

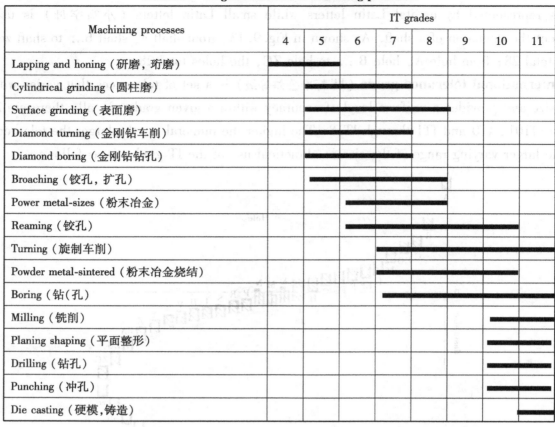

Tolerance zone(公差带) is the area bound by upper deviation and lower deviation as shown in Fig. 9.16. A tolerance zone consists of a fundamental deviation and an IT grade.

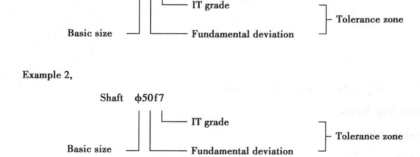

Tolerance value of a shaft or hole may be obtained by checking Tables 30 and 30 in Appendix 2 according to the basic size and its tolerance zone.

For example, according to the $\phi50H8$, from Table 31 in Appendix 2, it is known that upper deviation is $+0.039$, lower deviation 0.000. So, tolerance $\gamma = +0.039 - 0.000 = 0.0039$. Similarly, from $\phi50f7$ and Table 30, it is known that upper deviation is -0.025, lower deviation -0.050. So, tolerance $\gamma = -0.025 - (-0.050) = 0.025$.

Tolerance zone drawing(公差带图) is a sketch map expressing tolerance zone. Fig. 9.16 illustrates the tolerance zone drawings for tolerance dimensions $\phi50H8$ and $\phi50f7$.

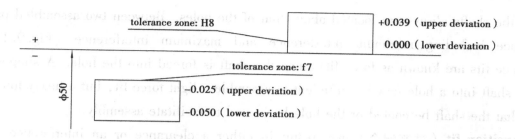

Fig. 9.16 Tolerance zone drawings.

②**Representing tolerance on dimensions**

Three methods of specifying tolerance on dimensions are as follows.

The basic size is followed by tolerance zone, $\phi 50H8$.

The basic size is followed by upper deviation and lower deviation, $\phi 50^{+0.039}_{0.000}$.

The basic size is followed by tolerance zone and upper deviation and lower deviation, $\phi 50H8(^{+0.039}_{0.000})$.

9.4.2.2 Fit (配合)

Fit is the general term used to signify the tightness or looseness that may result from the application of tolerances of the same basic size dimension of mating parts. The fit may be either a clearance fit, an interference fit, or a transition fit.

①**Terms**

Clearance fit (间隙配合) results in a clearance between two assembled parts under all tolerance conditions. In clearance fit, the shaft is always smaller than the hole, so the tolerance zones of the shafts always are located below that of the holes. Between two assembled parts exists clearance including maximum clearance and minimum clearance even zero clearance (Fig. 9.17 (a)). In practice, one part fits easily into another.

Interference fit (过盈配合) results in an interference between two assembled parts under all tolerance conditions. In interference fit, the shaft is always larger than the hole, so the tolerance

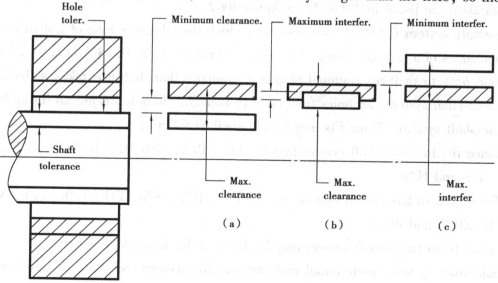

Fig. 9.17 Fits.

zones of the shafts always are located above that of the holes. Between two assembled parts exist interference including minimum interference and maximum interference (Fig. 9.17 (c)). Interference fits are known as force fits since the shaft is forced into the hole. A simple tap (轻敲) of a shaft into a hole may be sufficient to assemble a light force fit, but a heavy force fit may require that the shaft be cooled or the hole be heated to facilitate assembly.

Transition fit (过渡配合) may result in either a clearance or an interference under all tolerance conditions. In transition fit, the shaft may be either smaller or larger than the hole, so their tolerance zones appear superposition (Fig. 9.17 (b)). In practice, it is easy that either the shaft fits into the hole or the shaft is forced into the hole since in the transition fit, either clearance amount or interference amount are small.

②**Fit systems** (配合制度)

Chinese National Standard of Technology Drawing stipulates that two kinds of fit systems which are basic-hole system and basic-shaft system.

Basic-hole system (基孔制) is a system in which the tolerance zone of a hole is fixed while the tolerance zones of a shaft are changed to achieve three fits (Fig. 9.18(a)). In 28 holes from hole A, hole B... to hole ZC, National Standard stipulates that hole H is taken as basic hole (基准孔) in basic-hole system. 28 shafts from shaft a, shaft b... to shaft zc are all fitting shafts (配合轴) in basic hole-system. Three fits are as follows:

Clearance fits in basic-hole system may be H/a, H/b, H/c, H/cd, H/d, H/e, H/ef, H/f, H/fg, H/g and H/h.

Interference fit in basic-hole system may be H/p, H/r, H/s, H/t, H/u, H/v, H/x, H/y, H/z, H/za, H/zb and H/zc.

Transition fit in basic-hole system may be H/j, H/js, H/k, H/m, H/n.

In basic-hole system, preferential and common fits (优先常用配合) recommended by National Standard are listed in Table 28 in Appendix 2.

Basic-shaft system (基轴制) is a system in which the tolerance zone of a shaft is fixed while the tolerance zones of a hole are changed to achieve three fits (Fig. 9.18 (b)). In 28 shafts from shaft a, shaft b ... to shaft zc, National Standard stipulates that shaft h is taken as basic shaft (基准轴) in basic-shaft system. 28 holes from hole A, hole B... to hole ZC are all fitting holes (配合孔) in basic-shaft system. Three fits may be obtained as follows:

Clearance fits in basic-shaft system may be A/h, B/h, C/h, CD/h, D/h, E/h, EF/h, F/h, FG/h, G/h and H/h.

Interference fit in basic-shaft system may be P/h, R/h, S/h, T/h, U/h, V/h, X/h, Y/h, Z/h, ZA/h, ZB/h and ZC/h.

Transition fit in basic-shaft system may be JS/h, J/h, K/h, M/h, N/h.

In basic-shaft system, preferential and common fits recommended by National Standard are listed in Table 29 in Appendix 2.

③**Labeling fit dimensions**

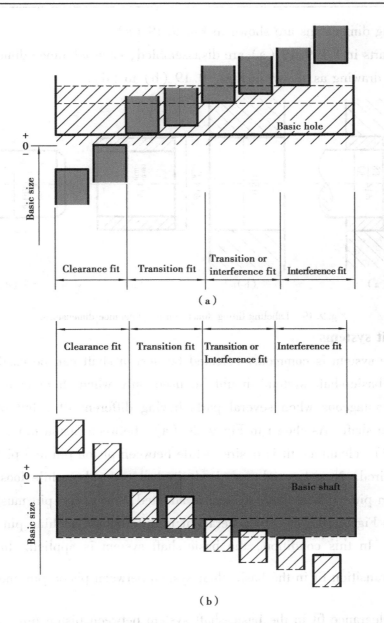

Fig. 9.18 Basic-hole system and basic-shaft system.

Writing fitting dimensions are as follows:

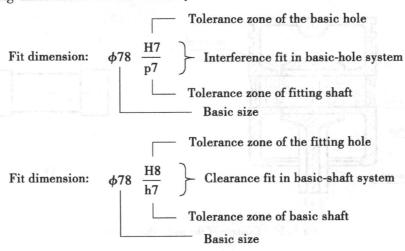

Labeling fitting dimensions are shown in Fig. 9.19 (a)

When three parts in Fig. 9.19 (a) are disassembled, each tolerance dimension is labeled on corresponding part drawing as shown in Figs. 9.19 (b) to (d).

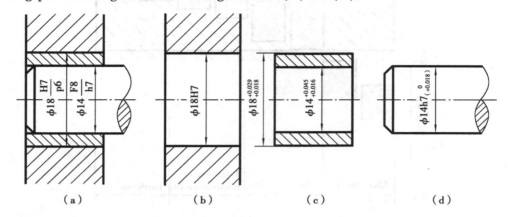

Fig. 9.19 Labeling fitting dimension and tolerance dimensions.

④Selecting fit systems

The basic-hole system is commonly selected because a shaft can be easily machined to any size desired. The basic-shaft system should be used only when there is a reason for it. For example, it is advantageous when several parts having different fits, but one nominal size is required on a single shaft. As shown in Fig. 9.20 (a), between piston pin (活塞销) and sleeve of piston-rod (连杆), clearance fit is desired while between piston pin and piston holes (活塞孔) transition fit is desired. Namely, tightness is desired at two sides while looseness is desired at middle of the piston pin. If basic-hole system is adopted, the piston pin must be machined into the shape shown in Fig. 9.18 (b). In general, it is different to machine pin because the pin is cold-finished shaft. In this condition, the basic-shaft system is applied. In Fig. 9.20 (a), $\phi 20 \frac{N6}{h5}$ indicates transition fit in the basic-shaft system between piston pin and piston holes while $\phi 20 \frac{N6}{h5}$ indicates clearance fit in the basic-shaft system between piston pin and sleeve of piston-rod.

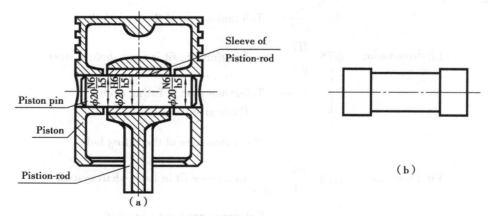

Fig. 9.20 Example of the basic-shaft system.

9.5 Reading Detail Drawings(阅读零件图)

9.5.1 Steps and procedures in reading detail drawing

The approaches and procedures in reading detail drawing are as follows.

9.5.1.1 Reading title block

First, understand the name of the part from title block to judge types of the part and deduce (推断) the complexity of the part. Secondly, understand the drawing scale to judge actual size of the part. If drawing scale is 1∶1, the part is as large as figures. If drawing scale is reduction scale such as 1∶2, 1∶5, the part is larger than figures. If drawing scale is enlarger scale such as 2∶1, 5∶1, the part is smaller than figures. Thirdly, understand material of the part to know surface quality approximately.

9.5.1.2 Analysis views

This is the core (核心) step in reading detail drawings. The analysis shape method (形体分析法) and the analyzing line and plane method (线面分析法) are the basic methods used in this step. Principle views including sections offer understanding of the interior and exterior structures of the part. Partial views, auxiliary views, cross-sections and zoom views are further helpful for understanding some partial structures.

9.5.1.3 Analyzing dimension and technical requirements

Analyze the dimension datums (尺寸基准) in width, height and depth. Find out the size dimensions and location dimensions for each component. Understand technical requirements such as surface roughness, dimension tolerance (尺寸公差) and other technical requirements written at the right lower corner of a detail drawing, such as requirement of hot treatment as well as radius of casting fillet and so on.

9.5.2 Example of reading detail drawing

Take the detail drawing of a bending arm (弯臂) as example to illustrate the procedure of reading a detail drawing (Fig. 9.21).

A glimpse on the title block informs us that the name of the part is bending arm, the material of the part is cast iron, and the drawing scale is 1∶2. The part belongs to fork and bracket-like parts. It may be used as connector and drivers in machine. Its blank (毛坯) is formed by casting and the real object is twice as large as the drawings.

Then from the drawings we learn that there are four figures in the detail drawing, two of which are principle views, one removed cross-section and one auxiliary view. The front view and top view show that the part consists of a bending arm and two sleeves chiefly. The left sleeve is big while the right one is small. The removed cross-section shows that the cross section of bending arm is elliptical shape. The bending arm extends from the front-side of big sleeve to right-upper to

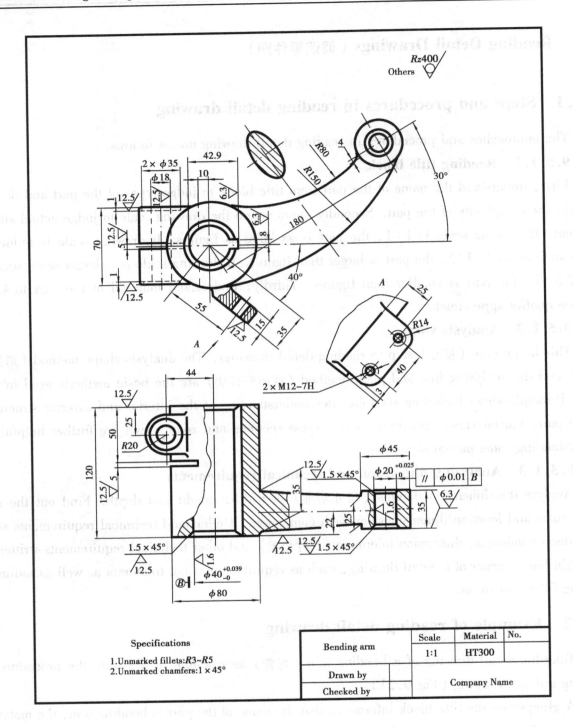

Fig. 9.21 Detail drawing of bending arm.

connect the small sleeve.

From the front view, the dimension datum in width direction may be found out, which is the axis of the large sleeve. Based on it, many location dimensions such as 10, 180, 44 (in the top view) are dimensioned. Again, from the front view, the dimension datum in height direction also may be found out, which is the axis of the large sleeve. Based on it, many location dimensions such as 5, 8, 30° and 40° are dimensioned. From the top view, the dimension datum in depth direction can be found out, which is rear side of the large sleeve. Based on it, many location dimensions such as 25, 50, 120 are dimensioned. Note that the front side of the large sleeve and the front-rear symmetric plane of bending arm are respectively secondary datums (辅助基准) in depth direction. Based on it, many location dimensions such as 22, 35 are dimensioned.

From the front and top views as well as removed cross-section, it may be deduced that the size of the cross section of the bending arm decreases gradually from left to right. Namely, the size of each cross section of arm is different.

Bosses (凸台) at the left-back side of big sleeve are used to clamp (夹紧). The height of the two bosses is 70 mm. There is a gap between the two bosses, in which the interval is 5 mm. Dimension $R20$ is size dimension in the two bosses. The location dimensions of the two bosses, 44 and 25, are dimensioned in the top view. In order to achieve clamp function, a group bolt joining is applied in assemble. Two countersink holes are machined to put the bolt' head or washer, respectively. Size dimensions of the two sunk holes are $2 \times \phi 35$, $\phi 18$, 1.

In order to meet the demand for locking large sleeve, two rifts (口子) are cut at the left-rear side of the large sleeve. One of the two rifts is perpendicular to axis of the large sleeve while the other is parallel to the axis of the large sleeve. Location dimensions of the rift perpendicular to the axis are 50 and 10 while size dimension is 5. Size dimension of the rift parallel to the axis is 5.

The broken-out section located in the front view reveals the threaded hole in the fixing plate whose true shape is represented in partial view A. There are two threaded holes in the fixing plate. Location dimensions about fixing plate are 55, 40° and 3 while dimensions 15, $R14$ are size dimensions. Location dimensions of the two threaded holes are 40 and 555 while dimension $2 \times M12$—7H is thread note of the threaded holes.

Two broken-out sections located in the top view reveal through holes in the large sleeve and small sleeve. Chamfer dimensions $1.5 \times 45°$ are used to front and rear sides of the two sleeves respectively. The top view and front view shows that there is a keyseat at the right side of inner wall of the big sleeve.

Two dimension tolerances are $\phi 40^{+0.039}_{0}$ and $\phi 20^{+0.025}_{0}$. Dimension $\phi 40^{+0.039}_{0}$ illustrates the inner diameter of big sleeve, which indicates the largest inner diameter may be made into 40.039 mm while the minimum diameter may be made into 40.000 mm. Tolerance is 0.039 mm. Dimension $\phi 20^{+0.025}_{0}$ illustrates the inner diameter of the small sleeve, which indicates the maximum diameter may be 20.025 mm and the minimum diameter may be 20.000 mm. Tolerance is 0.025.

The highest requirement for surface roughness is $\sqrt{1.6}$ for the inner hole of the big sleeve and small sleeve, the next surface roughness are $\sqrt{6.3}$, $\sqrt{12.5}$ and the rest surfaces are casting rough surfaces, which have roughness code $\sqrt{\ }$.

Fundamentals of Engineering Drawing

Fig. 9.22 Pictrial drawing of bending arm.

Fig. 9.22 is the pictorial drawing of the bending arm.

Chapter 10　Introduction of Assembly Drawings

After being made according to the specifications of the detail drawing, the parts can be assembled in accordance with directions of an assembly drawing. So, an assembly drawing is a working drawing that exhibits the working principle of a machine or components（部件）, assembled relationship and the shape of main part（主体零件的形状）. An assembly drawing embodies and communicates the designer's idea, moreover, with which workers can assemble, check, install, debug（调试）, use and maintain the product.

10.1　Contents

Fig. 10.1 is the assembly drawing of a sliding bearing（滑动轴承）and Fig. 10.2 is its pictorial drawing. Fig. 10.1 gives out a complete assembly drawing including the following contents:

Figures A group of figures that indicate the working principle, assembled relationship and the shape of main part of a machine or components.

Necessary Dimensions Only certain dimensions are needed for an assembly drawing. This is different from a detail drawing, where all dimensions are necessary for manufacturing and inspecting a part.

Technical requirements Technical requirements include the requirements for assembling, inspecting, installation, debugging and operation. Usually, those requirements are written by texts at the right-bottom of the assembly drawing.

Part list（明细栏）**and Title block.** Above the title block, there is a part list which gives out the part number, name, quantity, material. In general, in an assemble drawing, the parts are listed in the order of their importance, with the larger parts first and the standard parts such as screws, pins, etc, at the end.

10.2　Representations of Assembly Drawings

10.2.1　General representation

Multi-side orthographic projections often occur in assembly drawings. Some representation methods such as views, sections may be applied in an assembly drawing. But because there are many parts in an assembly drawing, superposition or coincident of lines, which makes assembly drawing unclear, may occur. So it is necessary to apply some prescriptive drawing and special representing methods for an assembly drawing.

10.2.2 Prescriptive drawing of assembly drawings (装配图的规定画法)

(1) The contacting surface of two parts is drawn in one line, but for non-contacting surface, it is drawn in two lines even the gap between them is very small (Fig. 10.3 (a)).

(2) In the section, the section lines of two adjacent parts should be different but the section lines of one part should be the same in all views (Fig. 10.3 (b)).

(3) For some objects such as shaft bolt, nut, screw, stud and washer etc., when they are cut longitudinally (纵切), these parts are not sectioned. However, they are cut transversely (横切地), these parts are sectioned.

As shown in Fig. 10.3 (c), the taper pin and shaft are not sectioned, but in order to represent the joining of the shaft and taper pin, a broken-out section is adopted at the right side of the shaft. As shown in Fig. 10.3 (d), section A-A is obtained by passing through the object transversely.

10.2.3 Special representation of assembly drawings (装配图的特殊画法)

(1) **Disassembly drawing** (拆卸画法)

In drawing certain view of an assembly drawing, if parts are hidden behind another part, then imagine the part is removed and draw the view again. The view is the disassembly drawing. As shown as Fig. 10.1, A-A half section is an example of disassembly drawing, in which part 9 is removed.

(2) **Cutting along conjoint plane between two parts** (沿结合面剖切)

In order to represent internal assembled relationship, suppose cut along the conjoint plane of two parts and then draw section as shown as Fig. 10.1 B-B half section. Note that do not draw section lines at the conjoint plane, but if certain part is cut off then draw section lines at the cross section of the part, e.g. the bolt shown in Fig. 10.1 B-B.

(3) **Imaginary drawing** (假象画法)

To represent the relationship between an assembling component and adjacent part, draw the adjacent part with imaginary line, e.g. the shaft in Fig. 10.1 A-A. Note that the shaft does not belong to the assembly body (装配体).

(4) **Amplified drawing** (放大画法)

In assembly drawing, tiny parts or gaps are drawn in enlarger scale in order to represent clearly.

(5) **Conventional practices** (简化画法)

In order to enhance drawing efficiency and reduce repeat, National Standard of Technology Drawing stipulates conventional practices (GB/T 16675.1—1996). For example, in drawing the thread fastener component (螺纹连接组件), only one group is drawn completely but other groups, only the position center lines are drawn.

In assembly drawing, bolt head and nut are drawn by simplified drawing in which chamfers are omitted, e.g. the nuts shown in the front view in Fig. 10.1.

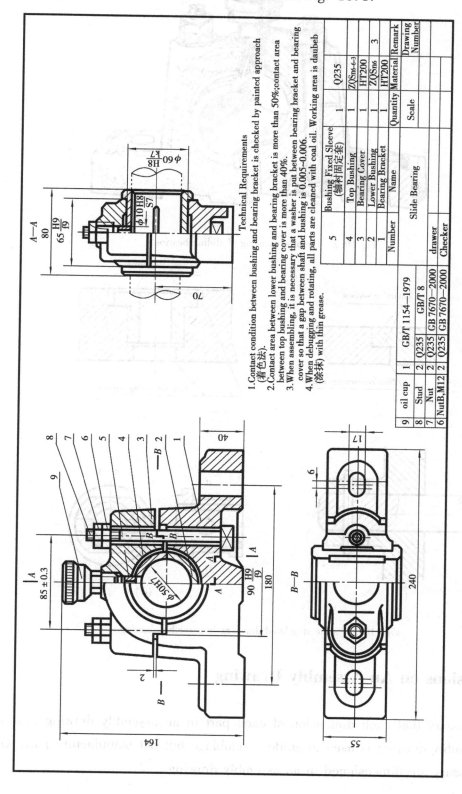

Fig.10.1 Assembly drawing of slide bearing(滑动轴承).

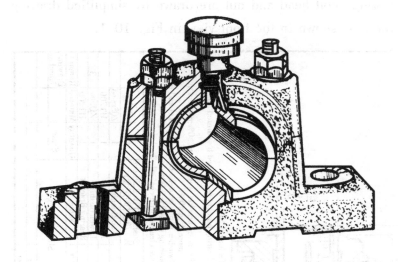

Fig. 10.2   Pictorial drawing of sliding bearing.

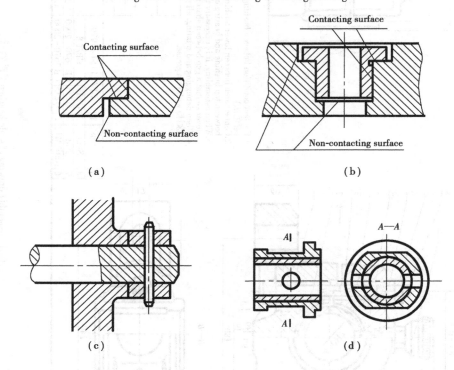

Fig. 10.3   Some stipulated drawings in assembly drawing.

10.3   Dimensions on An Assembly Drawing

It is unnecessary that each dimension of each part in an assembly drawing is dimensioned because the assembly drawing is used to guide assemblage but not manufacture part. Only those dimensions necessary are dimensioned in an assembly drawing.

10.3.1   **Performance dimensions**（性能尺寸）

Dimensions that satisfy usage requirement are called performance dimensions. As shown as

Fig. 10.1, the diameter of shaft supported by sliding bearing is $\phi 60$ which is a performance dimension.

10.3.2 Fit dimensions（配合尺寸）

Dimension denoting fitting relationship between parts is called fit dimension. As shown as Fig. 10.1, dimensions $\phi 60H8/k7$ and $65H9/k7$ are fit dimensions.

10.3.3 Installation dimensions（安装尺寸）

The dimensions needed for installing the assembly parts to baseplate（基础板）are called installation dimensions. As shown as Fig. 10.1, dimensions 180 and 17 are installation dimensions.

10.3.4 Total dimensions（总体尺寸）

The dimensions representing total width, depth and height are called total dimensions. They offer the data for packing, transportation and installation. As shown in Fig. 10.1, dimensions 240, 80 and 164 are total dimensions.

10.3.5 Other important dimensions（其他重要尺寸）

Besides the aforesaid four kinds of dimensions, there are other important dimensions in an assembly drawing. For example the limits dimensions of motive part, the position dimensions of core part（核心零件）, the important dimensions of main part（主体零件）, the dimensions guaranteed by installing and so on. As shown as Fig. 10.1, dimensions 55（the width of bearing bracket 轴承座）, 40（the height of bearing bracket 轴承座）, 70（the position dimension of the shaft supported by sliding bearing）and 2（the dimension of gap guaranteed by installing）belong to this category.

10.4 Part Numbered and Part List（零件序号和零件表）

Because there are many parts in an assembly drawing, all parts are numbered to designate the different parts. Leader line（指引线）goes from the inside of each part but turns to horizontal at the end. Thus part numbers are located above horizontal end, e.g. the number 1, number 2... shown in Fig. 10.1. Part numbers are filled in part list from bottom to top, where more information about each part is given, such as name of part, amount of parts, material of part and so on.

Note that part is numbered in turn clockwise or counterclockwise and they are arranged as close as possible（尽可能集中排列）. As shown in Fig. 10.1, all parts are numbered in turn with counterclockwise and they are arranged nearby front view.

Note that leader lines are not allowed to intersect mutually, don't forget to make a dot at its end in the part.

Appendix 1
Glossary（术语表）

1.1　Shop Processes

An engineer must be thoroughly familiar with the fundamental shop processes before he is qualified to preparing drawings that will fulfill the requirements of the production shops. In preparing working drawings, he must consider each and every individual process involved in the production of a piece and then specify the processes in terms that the shopman will undersdand. Althrough an acctuate knowledge of shop processes can be accquired only through actual experience in the various shop, it is possible for an apprentice to obtain a working knowledge of the fundamental oprations through study and observation. This appendix presents and explains the principal operations in the pattern shop（制模车间）, foundry（翻砂车间）, forge shop（锻工车间）, and machine shop（机械加工车间）.

1.2　Sand Casting（沙铸造）

Casting are formed by pouring molten metal into a mold or cavity（空穴）. In sand molding, the molten metal assumes（呈现）the shape of the cavity that has been formed in a sand mold by ramming（捣击）a prepared moist sand around a pattern and then removing the pattern. Although a casting shrinks somewhat in cooling, the metal hardens in the exact shape of the pattern used (Fig. 1.1).

A sand moled consists of at least two sections. The upper section, called the cope（上型箱）, and the lower section, called the drug（下型箱）, together form a box-shaped structure called a flask（砂箱）.

When large holes (19 mm and over) or interior passageways and openings（空隙）are needed in a casting, dry sand cores（干砂型心）are placed in the cavity. Cores exclude（排除）the metal from the space they ocupy and thus form desired openings. Large holes are cored to avoid an unnecessary boring operation（镗孔操作）.

The molder（铸工）when making a mold inserts in the sand a sprue stick（浇棒）that the removes after the cope has been rammed. This resulting hole, known as the sprue（注入口, 浇口）, conducts the motten matal to the gate, which is a passageway cut to the cavity. The adjacent hole called the riser（冒口）, provides an outlet（出口）for excess metal（盈余金属液体）.

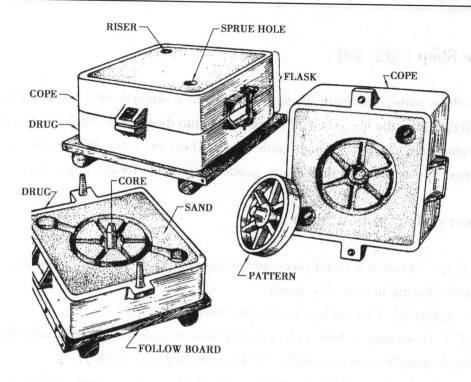

Fig. 1.1 Sand mold.

1.3 Pattern Shop (制模车间)

The pattern shop prepars patterns of all pieces that the foundry is to cast. A pattern usually is first constructed of light, strong wood, such as white pine or mahogeny, which, if only a few castings are required, may be used in making sand mold. In quantity production, however, where a pattern must be used repeatedly, the wooden one will not hold up, so a metal pattern (aluminum, brass, and so on) is made from it and is used in its place.

Every pattern must be constructed in such a way that it can be withdrawn (退出) from each section of the sand mold. If the pattern consists of two halves (split 分离的), the plane of separation should be so located that it will coincide with the plane of separation of the cope and the drag (Fig. 1.1). each portion of the pattern must be slightly tapered, so that it can be withdrawn without leaving a damaged cavity. Ordinarily, it is not necessary to specify the slight taper, known as draft (出模斜度).

Exterior corners are rounded for appearance and for the comfort of persons who must handle the part. A rounded internal corner is known as a round (圆角). A "filled-in" interior corner on a casting is called a fillet (内圆角). Sharp interior corners are avoided for two reasons: they are difficult to cast and they are likely to be potential points of failure (故障) because the crystals (晶体) of the cooling metal arrange themselves at a sharp corner in a weak pattern.

1.4　Forge Shop（锻工车间）

Many machine parts, especially those that must have strength and yet be light, are forged into shape（锻造成形）, the heated metal being forced into dies（冲垫，冲模）with a drop hammer（落锤）. Forgings are made of a high-grade steel. Dies are made by diemakers. Generally, special drawings, giving only the dimensions needed, are made for the forge shop.

1.5　Glossary of Common Shop

Anneal（v. 退火）To soften a metal piece and remove internal stresses by heating to its critical temperature and allowing to cool very slowly.

Arc-weld（v. 电弧焊接）To weld by electric-arc process.

Bore（v. 膛孔）To enlarge a hole with a boring tool in order to make it smooth, round and coaxial. Boring is usually done on a lathe（车床）or boring mill（镗床）.

Boss（n. 凸台）A circular projection, which is raised above a principal surface of a casting or forging.

Braze（v. 锡焊接）To join two pieces of metal by the use of hard solder. The solder is usually a copper-zinc alloy（铜锡合金）.

Broach（v. 凿孔）To machine a hole to a desired shape, uauslly other than（不同于）round. A tool with serrated（锯齿状的）edges is pushed or pulled through a hole to enlarge it to a required shape.

Burnish（v. 磨光）To smooth or apply a brilliant finish

Carburize（v. 渗碳）To harden the surface of a piece of low-grated steel by heating in a carbonizing（碳化）material to increase the carbon content and then quenching（淬火）.

Case-harden（v. 表面硬化）To harden a surface as described above or through the use of potassium cyanide（氰化钾）.

Chamfer（v. 倒角）To bevel（使成斜角）an external edge or corner.

Chase（v. 车螺丝）To cut threads on a lath using a chaser（螺纹梳刀）, a tool shaped to the profile of a thread.

Chill（v. 冷淬）To harden the surface of cast iron by sudden cooling against a metal molde.

Chip（v. 刨削）To cut away or remove surface defects（缺陷）with a chisel（凿子）.

Collar（n. 轴环）A cylindrical part fitted on（装上）a shaft to prevent a sliding movement.

Cold-work（v. 冷作）To deform metal stock by hammering, forming, drawing（拉拔）, etc., while a metal is at ordinary room temperature.

Core（v. 成芯）To form a hole or hollow cavity（洞）in a casting through the use of a core.

Counterbore（n. 钻沉孔，沉孔）To enlarge a hole to a given depth. 1. The cylindrical enlargement of the end of a drilled or bored hole. 2. A cutting tool for counterboring, having a piloted end the size of the drilled hole.

Countersink (*n.* 钻埋头孔，锥形沉孔) To form a depression to fit the conic head of a screw or the thickness of a plate so that the face will be level with the surface. A conic tool for countersinking.

Die casting (*n.* 压铸) Every accurate and smooth casting made by pouring a molten alloy (or composition, as Bakelite) usually under pressure into a metal mold or die. Distinguished from a casting made in sand.

Die stamping (*n.* 模压) A piece, usually of sheet metal, formed or cut by a die.

Drill (钻孔) To sink a hole with a drill, usually a twist drill (麻花钻). (n.) A pointed cutting tool rotated under pressure.

Drop forging (*n.* 落锤锻造) A wrought (锻造的) piece formed hot between dies under a drop hammer, or by preasure.

Feather (*n.* 滑键) A flat sliding key, which permits a pulley to move long the shaft prallel to its axis.

Fillet (*n.* 圆角) A rounded filling of the internal angle between two srfaces.

Flange (*n.* 法兰盘) A projecting rim (轮缘) or edge for fastening or stiffening (增大稳定性).

Forge (锻造) To shape metal while hot and plastic by hammering or forcing process either by hand or by machine.

Galvanize (电镀) To treat with a bath of lead and zinc to prevent rusting (生锈).

Graduate (标刻度) To divide a scale or dial (刻度盘) into regular spaces.

Grind (磨削) To finish or polish a surface by means of an abrasive wheel (砂轮).

Harden (淬硬) To heat hardenable steel above critical temoerature and quench in bath.

Kerf (*n.* 切口) The channel or groove cut by a saw or other tool.

Key (*n.* 键) A piece used between a shaft and a hub to prevent the movement of one relative to the other.

Keyway or Keyseat (*n.* 键槽) A longitudinal groove cutin a shaft or a hub to receive a key. A key rests in a keyseat and slides in a keyway.

Knurl (*v.* 滚花) To roughen a cylindrical surface to produce a better grip (紧握) for the fingers.

Lap (*v.* 研磨) To finish or polish with a piece of soft metal, wood, or leather impregnated with an abrasive (磨料).

Lug (*n.* 耳状物,凸缘) A projection (凸起) or ear, which has been cast or forged as a portion of a piece to provide a support or to allow the attachment of another part.

Malleable Casting (*n.* 可锻铸造) A casting that has been annealed (退火) to toughen (强韧) it.

Mill (*v.* 轧制) To machine a piece on a milling machine by means of a rotating toothed cutter.

Neck (*v.* 颈状凹槽) To cut a circumferential groove round a shft.

Peen (*v.* 锤尖敲打) To stretch or bend over metal using the peen end (ball end) of a hammer.

Pickle (*v.* 蚀,浸,洗) To remove scale and rust from a casting or forging by immersing it in an acid that.

Plane (*v.* 刨平) To machine a flat surface on a planer, a machine having a fixed tool and a

Fundamentals of Engineering Drawing

reciprocating（往复）bed.

Polish（*v.* 抛光）To make a surface smooth and lustrous through the use of a fine abrasive.

Punch（*v.* 冲压）To perforate a thin piece of metal by shearing out a circular wad with a nonrotating tool under pressure.

Ream（*v.* 铰）To finish a hole to an exact size using a rotating fluted cutting tool known as a reamer.

Rib（*n.* 肋板）A thin component of a part that acts as a brace or support.

Rivet（*n.* 铆钉）A headed shank, which more or less permanently unites two pieces.（v）To fasten steel plates with rivets.

Round（*n.* 圆角）A rounded external corner on a casting.

Sandblast（*v.* 喷沙）To clean the surface of castings or forgings by means of sand forced from a nozzle at a high velocity.

Shear（*v.* 剪切）To cut off sheet or bar metal through the shearing action of two blades.

Shim（*n.* 衬垫）A thin metal plate, which is inserted between two surfaces for the purpose of adjustment.

Spotface（*v.* 锪平，刮孔口平面，锪孔）To finish a round spot on the rough surface of a casting at a drilled hole for the purpose of providing a smooth seat for a bolt or screw head.

Spot Weld（*v.* 点焊）To weld two overlapping metal sheets in spots by means of the heat of resistance to an electric current between a pair of electrodes.

Steel Casting（*n* 钢铸）A casting made of cast iron to which scrap steel has been added.

Swage［*v.* 型锻；（用陷型模）使成形］To form metal with a "swage block," a tool so constructed that through hammering or pressure the work may be made to take a desired shape.

Sweat（*v.* 熔焊）To solder together by clamping the pieces in contact with soft solder between and then heating.

Tack Weld（*n.* 间断焊）A weld of short intermittent sections.

Tap（*v.* 攻螺纹）To cut an internal thread, by hand or with power, by screwing into the hole a fluted tapered tool having thread-cutting edges.

Taper（*v.* 使逐渐变细；使变尖）To make gradually smaller toward one end.（*n.* 锥度）Gradual diminution of diameter or thickness of an elongated object.

Taper Pin（*n.* 锥形销）A peered pin used for fastening hubs or collars to shafts.

Temper（*v.* 回火）To reduce the hardness of a piece of hardened steel through reheating and sudden quenching（淬火）.

Template（*n.* 模板）A pattern cut to a desired shape, which is used in layout work to establish shearing lines, to locate holes, etc.

Turn（*v.* 车削）To turn-down or machine a cylindrical surface on a lathe.

Undercut（*n.* 底切）A recessed（使凹进的）cut.

Upset（*v.* 镦粗）To increase the diameter or form a shoulder on a piece during forging.

Weld（*v.* 焊接）To join two pieces of metal by pressure or hammering after heating to the fusion point.

Appendix 2

1. Threads

1.1 Metric Threads (GB 192—1981, GB 193—1981)

1.1.1 Basic thread form

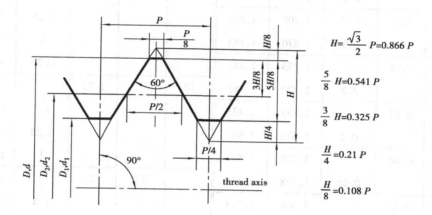

$H = \dfrac{\sqrt{3}}{2} P = 0.866 P$

$\dfrac{5}{8} H = 0.541 P$

$\dfrac{3}{8} H = 0.325 P$

$\dfrac{H}{4} = 0.21 P$

$\dfrac{H}{8} = 0.108 P$

Fig. 1

D—the major diameter of internal thread
d—the major diameter of external thread
P—Pitch
D_2—the pitch diameter of internal thread
d_2—the pitch diameter of external thread
H—the height of original angle
D_1—the minor diameter of internal thread
d_1—the minor diameter of external thread
Nominal diameter—the major diameter of both internal and external thread

Label Example

①Coarse metric thread, nominal diam. $D = 24$:

M24

②Fine metric thread, Nominal diam. $D = 24$mm, Pitch, $P = 2$mm

M24 × 2

③Fine metric thread, Nominal diam. $D = 24$, Pitch, $P = 1.5$, Left hand:

M24 × 1.5 LH

1.1.2 Specification of metric thread (Table 1)

Table 1 Metric threads (GB/T 1982—1981) (mm)

Nominal diam. D, d 1st series	Nominal diam. D, d 2nd series	Pitches P Coarse	Pitches P Fine	Pitch diam. D_2 or d_2	Minor diam. D_1 or d_1	Nominal diam. D, d 1st series	Nominal diam. D, d 2nd series	Pitches P Coarse	Pitches P Fine	Pitch diam. D_2 or d_2	Minor diam. D_1 or d_1
1		0.25		0.838	0.729	6		1		5.350	4.917
			0.2	0.870	0.783				0.75	5.513	5.188
	1.1	0.25		0.938	0.829	8		1.25		7.188	6.647
			0.2	0.970	0.883				1	7.350	6.917
1.2		0.25		1.038	0.929				0.75	7.513	7.188
			0.2	1.070	0.983	10		1.5		9.026	8.376
	1.4	0.3		1.205	1.075				1.25	9.188	8.647
			0.2	1.270	1.183				1	9.350	8.917
1.6		0.35		1.373	1.221				0.75	9.513	9.188
			0.2	1.470	1.383	12		1.75		10.863	10.106
	1.8	0.35		1.573	1.421				1.5	11.026	10.376
			0.2	1.670	1.583				1.25	11.188	10.647
2		0.40		1.740	1.567				1	11.350	10.917
			0.25	1.838	1.729		14	2		12.701	11.835
	2.2	0.45		1.908	1.713				1.5	13.026	12.376
			0.25	2.038	1.929				1	13.350	12.917
2.5		0.45		2.208	2.013	16		2		14.701	13.835
			0.35	2.273	2.121				1.5	15.026	14.376
3		0.50		2.675	2.459				1	15.350	14.917
			0.35	2.773	2.621		18	2.5		16.376	15.294
	3.5		0.35	3.273	3.121				2	16.701	15.835
4		0.7		3.545	3.242				1.5	17.206	16.376
									1	17.350	16.917
			0.5	3.675	3.459	20		2.5		18.376	17.294
	4.5		0.5	4.175	3.959				2	18.701	17.835
5		0.8		4.480	4.134				1.5	19.026	18.376
			0.5	4.675	4.459				1	19.350	18.917

(Continued)

Nominal diam. D, d		Pitches P		Pitch diam. D_2 or d_2	Minor diam. D_1 or d_1	Nominal diam. D, d		Pitches P		Pitch diam. D_2 or d_2	Minor diam. D_1 or d_1
1st series	2nd series	Coarse	Fine			1st series	2nd series	Coarse	Fine		
	22	2.5		20.376	19.294		39	4		36.402	34.670
			2	20.701	19.835				3	37.051	35.753
			1.5	21.026	20.376				2	37.701	36.835
			1	21.350	20.917				1.5	38.026	37.376
24		3		22.051	20.751	42		4.5		39.077	37.129
			2	22.701	21.835				3	40.015	38.753
			1.5	23.026	22.376				2	40.701	39.835
			1	23.350	22.197				1.5	41.026	40.376
	27	3		25.051	23.752		45	4.5		42.077	40.129
			2	25.701	24.835				3	43.051	41.752
			1.5	26.026	25.376				2	43.701	42.835
			1	26.350	25.917				1.5	45.026	43.376
30		3.5		27.727	26.211	48		5		44.752	42.587
			2	28.701	27.835				3	46.051	44.752
			1.5	29.026	28.376				2	46.701	45.835
			1	29.350	28.917				1.5	47.026	46.376
	33	3.5		30.727	29.211		52	5		48.752	46.587
			2	31.701	30.835				3	50.051	48.752
			1.5	32.026	31.376				2	50.701	49.835
36		4		33.402	31.620				1.5	51.026	50.376
			3	34.051	32.752	56		5.5		52.428	50.046
			2	34.701	33.835						
			1.5	35.026	34.376				4	53.402	51.670

Appendix 2

237

1.2 Trapezoid Threads (GB/T 5796.1—5796.4—1986)

1.2.1 Basic thread form

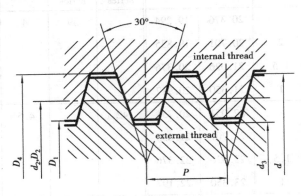

Fig. 2

d—the major diameter (nominal diameter)

P—Pitch

D_1—the minor diameter of internal thread

d_2—the pitch diameter of external thread

D_2—the pitch diameter of internal thread

d_3—the pitch diameter of external thread

d_4—the major diameter of internal thread

Lable Examlpe

External thread, nominal diameter 40mm, pitch 7 mm, double thread, left, tolerance range code 7e, normal engagement

Tr40×14(P7)LH-7e

1.2.2 Specification of trapezoid threads (Tables 2 to 3)

Table 2 Series of diameters and pitches (GB 5796.2—1986) (mm)

Nominal diam.		Pitches P		Nominal diam.		Pitches P	
1st Series	2nd Series	Priority select	Ordinary	1st Series	2nd Series	Priority select	Ordinary
10		2	1.5		22	5	3;8
	11	2	3	24		5	3;8
12		3	2		26	5	3;8
	14	3	2	28		5	3;8
16		4	2		30	6	3;10
	18	4	2	32		6	3;10
20		4	2		34	6	3;10

(Continiued)

Nominal diam.		Pitches P		Nominal diam.		Pitches P	
1st Series	2nd Series	Priority select	Ordinary	1st Series	2nd Series	Priority select	Ordinary
36		6	3;10	48		8	3;12
	38	7	3;10		50	8	3;12
40		7	3;10	52		8	3;12
	42	7	3;10		55	9	3;14
44		7	3;12	60		9	3;14
	46	8	3;12		65	10	3;16

Table 3 Trapezoid threads (GB 5796.3—1986)   (mm)

Pitch P	External thread		Pitch. Diameter d_2, D_2	Internal thread		Pitch P	External thread		Pitch. diam. d_2, D_2	Internal thread	
	Max. diam. d	Min. diam. d_3		Max. diam. D_4	Min. diam. d_3		Max. diam. d	Min. diam. d_3		Max. diam. D_4	Min. diam. d_3
1.5	10	8.20	9.25	10.30	8.50	3	48	44.50	46.50	48.50	45.00
2	10	7.59	9.00	10.50	8.00	3	50	46.50	48.50	50.50	47.00
2	11	8.50	10.00	11.50	9.00	3	52	48.50	50.50	52.50	49.00
2	12	9.50	11.00	12.50	10.00	3	55	51.50	53.50	55.50	52.00
2	14	11.50	13.00	14.50	12.00	3	60	56.50	58.50	60.50	57.00
2	16	13.50	15.00	16.50	14.00	4	16	11.50	14.00	16.50	12.00
2	18	15.50	17.00	18.50	16.00	4	18	13.50	16.00	18.50	14.00
2	20	17.50	19.00	20.50	18.00	4	20	15.50	18.00	20.50	16.00
3	11	7.50	9.50	11.50	8.00	4	65	60.50	63.00	65.50	61.00
3	12	8.50	10.50	12.50	9.00	5	22	16.50	19.50	22.50	17.00
3	14	10.50	12.50	14.50	11.00	5	24	18.50	21.50	24.50	19.00
3	22	18.50	20.50	22.50	19.00	5	26	20.50	23.50	26.50	21.00
3	24	20.50	22.50	24.50	21.00	5	28	22.50	25.50	28.50	23.00
3	26	22.50	24.50	26.50	23.00	6	30	23.50	27.50	31.00	24.00
3	28	24.50	26.50	28.50	25.00	6	32	25.00	29.00	33.00	26.00
3	30	26.50	28.50	30.50	27.00	6	34	27.00	31.00	35.00	28.00
3	32	28.50	30.50	32.50	29.00	6	36	29.00	33.00	37.00	30.00
3	34	30.50	32.50	34.50	31.00	7	38	30.00	34.50	39.00	31.00
3	36	32.50	34.50	36.50	33.00	7	40	32.00	36.50	41.00	33.00
3	38	34.50	36.50	38.50	35.00	7	42	34.00	38.50	43.00	35.00
3	40	36.50	38.50	40.50	37.00	7	44	36.00	40.50	45.00	37.00
3	42	38.50	40.50	42.50	39.00	8	22	13.00	18.00	23.00	14.00
3	44	40.50	42.50	44.50	41.00	8	24	15.00	20.00	25.00	16.00
3	46	42.50	44.50	46.50	43.00	8	26	17.00	22.00	27.00	18.00

(Continued)

Pitch P	External thread Max. diam. d	External thread Min. diam. d_3	Pitch. Diameter d_2, D_2	Internal thread Max. diam D_4	Internal thread Min. diam. d_3	Pitch P	External thread Max. diam. d	External thread Min. diam. d_3	Pitch. diam. d_2, D_2	Internal thread Max. diam. D_4	Internal thread Min. diam. d_3
8	28	19.00	24.00	29.00	20.00	10	38	27.00	33.00	39.00	28.00
8	46	37.00	42.00	47.00	38.00	10	40	29.00	35.00	41.00	30.00
8	48	39.00	44.00	49.00	40.00	10	65	54.00	60.00	66.00	55.00
8	50	41.00	46.00	51.00	42.00	12	44	31.00	38.00	45.00	32.00
8	52	43.00	48.00	53.00	44.00	12	46	33.00	40.00	47.00	34.00
9	55	45.00	50.00	56.00	46.00	12	48	35.00	42.00	49.00	36.00
9	60	50.00	55.50	61.00	51.00	12	50	37.00	44.00	51.00	38.00
10	30	19.00	25.00	31.00	20.00	12	52	39.00	46.00	53.00	40.00
10	32	21.00	27.00	33.00	22.00	14	55	39.00	48.00	57.00	41.00
10	34	23.00	29.00	35.00	24.00	14	60	44.00	53.00	62.00	46.00
10	36	25.00	31.00	37.00	26.00	16	65	47.00	57.00	67.00	49.00

1.3 Straight Pipe Threads (GB/T 7307—1987)

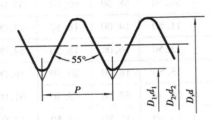

Fig. 3

Label Example

Internal Thread, Dimension code 1½: G 1½

B grade External Thread, Dimension code 1½: G 1½ B

(The cylindrical external thread sealed not by screw)

Table 4 Straight pipe threads (GB/T 7307—1987) (mm)

Dimension Code	The teeth Per 25.4 mm	Pitch P	Basic Dimension Max. diam.	Basic Dimension Pitch. diam.	Basic Dimension Min. diam.
1/8	28	0.907	9.728	9.147	8.566
1/4	19	1.337	13.157	12.301	11.445
3/8	19	1.814	16.662	15.806	14.950
1/2	14	1.814	20.955	19.793	18.631
5/8	14	1.814	22.991	21.749	20.587
3/4	14	1.814	26.441	25.279	24.117
7/8	14	1.814	30.201	29.039	27.877

(Continued)

Dimension Code	The teeth Per 25.4 mm	Pitch P	Basic Dimension		
			Max. diam.	Pitch. diam.	Min. diam.
1	11	2.309	33.249	31.770	30.291
1⅛		2.309	37.897	36.418	34.939
1¼		2.309	41.910	40.431	38.952
1½		2.309	47.803	46.324	44.845
1¾		2.309	53.746	52.267	50.788
2		2.309	59.614	58.135	56.656
2¼		2.309	65.710	64.231	62.752
2½		2.309	75.184	73.705	72.226
2¾		2.309	81.534	80.055	78.576
3		2.309	87.884	86.405	84.926

1.4 Taper Pipe Threads (GB/T 7306—1987)

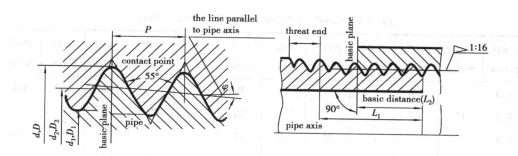

Fig. 4

Label Example

Taper external thread, dimension code 3/4: $R3/4$

Taper internal thread, dimension code 3/4: $R_c = 3/4$

Cylindrical internal thread, dimension code 1½: $R_p 1½$

Table 5 Taper pipe threads (GB/T 7306—1987)　　　　　(mm)

Nominal pipe size	Threads Per 25.4 mm	Pitch P	The Basic diameter of Basic plane			The basic distance L_2	The length of effective thread L_1
			Max. diam.	Pitch. diam.	Min. diam.		
1/8	28	0.907	9.728	9.147	8.566	4.0	6.5
1/4	19	1.337	13.157	12.301	11.445	6.0	9.7
3/8			16.662	15.806	14.950	6.4	10.0
1/2	14	1.814	20.955	19.793	18.631	8.2	13.2
3/4			26.441	25.279	20.587	9.5	14.5

Fundamentals of Engineering Drawing

(Continued)

Nominal pipe size	Threads Per 25.4 mm	Pitch P	The Basic diameter of Basic plane			The basic distance L_2	The length of effective thread L_1
			Max. diam.	Pitch. diam.	Min. diam.		
1	11	2.309	33.249	31.770	30.291	10.4	16.8
1¼			41.910	40.431	38.952	12.7	19.1
1½			47.803	46.324	44.845	12.7	19.1
2			59.614	58.135	56.656	15.9	23.4
2½			75.184	73.705	72.226	17.5	26.7
3			87.884	86.405	84.926	20.6	29.8
4			113.030	111.551	110.072	25.4	35.8

2. Standard Dimensions and Standard Structures

2.1 Standard Dimensions (GB/T 2822—1981)

Table 6   Standard dimensions (GB/T 2822—1981)   (mm)

Ra10 Series	Ra20 Series	Ra10 Series	Ra20 Series	Ra40 Series	Ra10 Series	Ra20 Series	Ra40 Series	Ra10 Series	Ra20 Series	Ra40 Series	Ra10 Series	Ra20 Series	Ra40 Series
2.0	2.0		11			28	28			67	160	160	160
	2.2	12	12	12			30		71	71			170
2.5	2.5			13	32	32	32			75		180	180
	2.5		14	14			34	80	80	80			190
3.0	3.0			15		36	36			85	200	200	200
	3.5	16	16	16			38			90			210
5.0	5.0			17	40	40	40			95		220	220
	5.5		18	18			42	100	100	100			240
4.0	4.0			19		45	45			105	250	250	250
	4.5	20	20	20			48		110	110			260
6.0	6.0			21	50	50	50			120		280	280
	7.0		22	22			53	125	125	125			300
8.0	8.0			24		56	56			130	320	320	320
	9.0	25	25	25			60			140			340
10.0	10.0			26	63	63	63			150		360	360

2.2 End Forms and Dimensions of Bolts and Screws (Including Chamfers) (GB/T 2—1985)

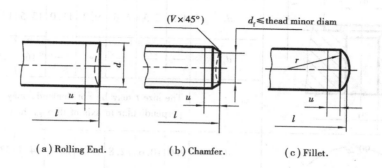

(a) Rolling End. (b) Chamfer. (c) Fillet.

Fig. 5

d—nominal diameter (the length of not complete thread)
d_1——the diameter of chamfer plane
P——Pitch
r—1.4d

Table 7 Chamfer sizes (mm)

d	2	2.5	3	3.6	4	4.5	5	6	8	10	12	14	16
d_f	1.5	2	2.4	2.9	3.2	3.7	4.1	4.9	6.4	8	10	11.5	13.5
V	0.25	0.25	0.3	0.3	0.4	0.4	0.45	0.55	0.8	1	1	1.25	1.25
d	18	20	22	24	27	30	33	36	39	42	45	48	52
d_f	15	17	19	20	23	26	29	31	34	37	40	42	46
V	1.5	1.5	1.5	2	2	2	2	2.5	2.5	2.5	2.5	3	3

Note: The chart is defined by GB 2-85.

2.3 Sunk Holes and Through Holes

Sizes of through Hole (GB/T 5277—1985)
Sunk holes (GB 152.2—88, GB/T 152.3—1988, GB/T 152.4—1988)

Table 8 Counterbores and through holes (mm)

The diam. of bolt or screw		4	5	6	8	10	12	16	20	24	30	36
The diam. of through hole (GB 5277—85)	Fine assembly (精装配)	4.3	5.3	6.4	8.4	10.5	13	17	21	25	31	37
	Middle assembly (中等装配)	4.5	5.5	6.6	9	11	13.5	17.5	22	26	33	39
	Coarse assembly (粗装配)	4.8	5.8	7	10	12	14.5	18.5	24	28	35	42

(Continued)

		The diam. of bolt or screw	4	5	6	8	10	12	16	20	24	30	36
Use for hex. head bolt and hex. head screw(六角头螺栓及螺钉)		d_2	10	11	13	18	22	26	33	40	48	61	71
		d_1	4.5	5.5	6.6	9.0	11.0	13.5	17.5	22.0	26	33	39
		d_3	—	—	—	—	16	20	24	28	36	42	
		t	The size t may be not demand, may be made into the plane perpendicular to axis of through hole.										
Use for flat head screw (用于沉头及半沉头螺钉)		d_2	9.6	10.6	12.8	17.6	20.3	24.4	32.4	40.4			
		d_1	4.5	5.5	6.6	9	11	13.5	17.5	22			
		$t\approx$	2.7	2.7	3.3	4.6	5.0	6.0	8.0	10.0			
Use for fillister head screw (用于圆柱头螺钉)		d_2	8	10	11	15	18	20	26	33			
		d_1	4.5	5.5	6.6	9.0	11.0	13.5	17.5	22.0			
		t	3.2	4.0	4.7	6.0	7.0	8.0	10.5	12.5			
Use for hex. socket head screw (用于圆柱头内六角螺钉)		d_2	8.0	10.0	11.0	15.0	18.0	20.0	26.0	33.0			
		d_1	4.5	5.5	6.6	9.0	11.0	13.5	17.5	22.0			
		t	4.6	5.7	6.8	9.0	11.0	13.0	17.5	21.5			

3. Standard Parts (标准件)

3.1 Screws

3.1.1 Fillister head screws (开槽圆柱头螺钉) (GB/T 65—2000)

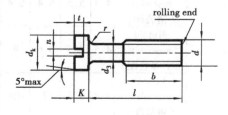

Fig. 6

Label Example

Nominal diam. $d = M5$, Nominal Length $L = 20$mm, Screw GB/T 65 M5 × 20

Table 9 Fillister head screws（开槽圆柱头螺钉）   (mm)

Nominal diam. (d)	M4	M5	M6	M8	M10
Pitch (P)	0.7	0.8	1	1.25	1.5
Max. (d_k)	7	8.5	10	13	16
Max. (k)	2.6	3.3	3.9	5	6
t	1.1	1.3	1.6	2	2.4
N	1.2	1.2	1.6	2	2.5
r	0.2	0.2	0.25	0.4	0.4
$\dfrac{l(s\tan dard\ range)}{b}$	$\dfrac{5\sim40}{b\approx l}$	$\dfrac{6\sim50}{38}$	$\dfrac{8\sim60}{38}$	$\dfrac{10\sim80}{38}$	$\dfrac{12\sim80}{38}$
L (Series)	5,6,8,10,12,(14),16,20,25,30,35,40,45,50,(55),60,(65),70,(75),80				

Note: ① Use the standard values in brackets as much as impossible.

② When the nominal length is in 40mm, full rod is made into thread.

3.1.2 Flat socket head screws（开槽沉头螺钉）(GB/T 68—2000),（开槽半沉头螺钉）(GB/T 69—2000)

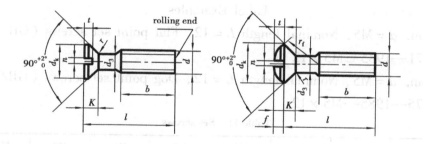

Fig. 7

$d_3 \approx$ Mid. diam. or $d_3 =$ Max. diam.

Label Examples

Nominal diam. $d = M5$, Nominal Length $L = 20$, Screw: GB/T 68—M5 × 20

Nominal diam. $d = M5$, Nominal Length $L = 20$, Screw: GB/T 69—M5 × 20

Table 10 Flat socket head screws (GB/T 68—2000) and (GB/T 69—2000)   (mm)

Nominal diam. (d)	M4	M5	M6	M8	M10
Pitch (P)	0.7	0.8	1	1.25	1.5
max. (d_k)	7	8.5	10	13	16
max. (K)	2.6	3.3	3.9	5	6
t	1.1	1.3	1.6	2	2.4
n	1.2	1.2	1.6	2	2.5

(Continued)

Nominal diam. (d)	M4	M5	M6	M8	M10
r	0.2	0.2	0.25	0.4	0.4
$\dfrac{L(\text{standard range})}{b}$	$\dfrac{5\sim40}{b\approx l}$	$\dfrac{6\sim50}{38}$	$\dfrac{8\sim60}{38}$	$\dfrac{10\sim80}{38}$	$\dfrac{12\sim80}{38}$
L(Series)	5,6,8,10,12,(14),16,20,25,30,35,40,45,50,(55),60,(65),70,(75),80				

Note: ①Use the standard values in brackets as much as impossible.

②When the nominal length is in 40 mm, full rod is made into thread.

3.1.3 Set screws

Flat point set screws（锥端紧定螺钉）（GB/T 71—1985）**and dog point set screws**（长圆柱端紧定螺钉）（GB/T 75—1985）

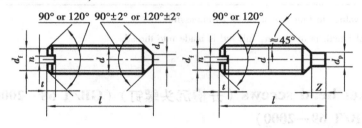

Fig. 8

Label Examples

Nominal diam. $d = \text{M5}$, Nominal length $L = 12$, Flat point set screws (GB/T 71—1985):
Set screw GB/T 71—1985—M5 × 12

Nominal diam. $d = \text{M5}$, Nominal length $L = 12$, Dog point set screws (GB/T 75—1985):
Set screw GB/T 75—1985—M5 × 12

Table 11 Set screws (mm)

Nominal diam. (d)	M12	M3	M4	M5	M6	M8	M10	M12
Pitch (P)	0.4	0.5	0.7	0.8	1	1.25	1.5	1.75
$d_f \approx$	Minor diameters							
d_t (GB/T 71—1985)	0.2	0.3	0.4	0.5	1.5	2	2.5	3
d_p (GB/T 75—1985)	1	2	2.5	3.5	4	5.5	7	8.5
n	0.25	0.4	0.6	0.8	1	1.2	1.6	2
t	0.84	1.05	1.42	1.63	2	2.5	3	3.6
Z (GB/T 75—198)	1	1.5	2	2.5	3	4	5	6
l range	3~10	5~16	6~20	8~25	8~30	10~40	12~50	14~60
L Series	3,4,5,6,8,10,12,(14),16,20,25,30,35,40,45,50,55,60							

Note: ①Use the standard values in brackets as much as impossible.

② When the nominal length is in 40mm, full rod is made into thread.

3.2 Bolts

3.2.1 Hex. bolts—C grade (GB/T 5780—2000)

3.2.2 Hex. bolts—Full thread—C grade (GB/T 5781—2000)

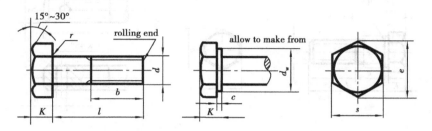

Fig. 9
Label Examples

Nominal diam. d = M12, Nominal length L = 80 mm, C grade hex.: Bolt GB/T 5780—M12×80

Nominal diam. d = M12, Nominal length L = 80 mm, full thread C grade: Bolt GB/T 5781—M15×80

Table 12 Hex. bolts (GB/T 5780—2000) and hex. bolt (GB/T 5781—2000) (mm)

Nominal diam. d		M5	M6	M8	M10	M12	M16	M20	M24	M30
s (max)		8	10	13	16	18	24	30	36	46
k		3.5	4	5.3	6.4	7.5	10	12.5	15	18.7
e (min)		8.63	10.89	14.20	17.59	19.89	26.17	32.95	39.55	50.85
c (max)		0.5	0.5	0.6	0.6	0.6	0.8	0.8	0.8	0.8
r		0.2	0.25	0.4	0.4	0.6	0.6	0.8	0.8	1.8
d_w (min)		6.7	8.7	11.4	14.4	16.4	22	27.7	33.2	42.7
a (max)		3.2	4	5	6	7	8	10	12	14
$\dfrac{l}{b}$	GB 578—86	$\dfrac{25\sim50}{16}$	$\dfrac{30\sim60}{18}$	$\dfrac{30\sim80}{22}$	$\dfrac{40\sim100}{26}$	$\dfrac{45\sim120}{30}$	$\dfrac{55\sim120}{38}$	$\dfrac{40\sim100}{26}$	$\dfrac{80\sim120}{54}$	$\dfrac{90\sim120}{66}$
							$\dfrac{130\sim160}{44}$	$\dfrac{130\sim200}{52}$	$\dfrac{130\sim200}{60}$	$\dfrac{130\sim200}{72}$
									$\dfrac{220\sim240}{73}$	$\dfrac{220\sim300}{85}$
	GB 5781—86	$\dfrac{10\sim40}{l-a}$	$\dfrac{12\sim50}{l-a}$	$\dfrac{16\sim65}{l-a}$	$\dfrac{20\sim80}{l-a}$	$\dfrac{25\sim100}{l-a}$	$\dfrac{35\sim100}{l-a}$	$\dfrac{40\sim100}{l-a}$	$\dfrac{50\sim100}{l-a}$	$\dfrac{60\sim100}{l-a}$
L series		10,12,16,20,25,30,35,40,45,50,(55),60,(65),70,80,90,100,110,120,130,140,150,160, 180,200,220,240,260,280,300								

Note: ① As impossible as using the standard in brackets.
② When the nominal length in 40 mm, full rod is made into thread.

3.2.3 Hex. bolts—A and B grade (GB/T 5782—2000)

3.2.4 Hex. bolts—Full thread—A and B grade (GB/T 5783—2000)

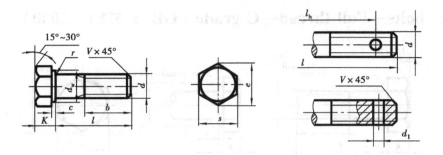

Fig. 10

Label Examples

Nominal diam. $d=12$, Nominal length $L=80$ mm, A grade hex. bolt:

Bolt GB/T 5782　M12×80

Nominal diam. $d=12$, Nominal length $L=80$ mm, full thread, A grade hex. bolt:

Bolt GB/T 5783　M12×80

Table 13　Hex. bolts (GB/T 5782—2000) and Hex. bolts (GB/T 5783—2000)　　(mm)

Nominal diam. d		M4	M5	M6	M8	M10
s		7	8	10	13	16
k		2.8	3.5	4	5.3	6.4
e (min)	Grade A	7.66	8.79	11.05	14.38	17.77
	Grade B	—	8.63	10.89	14.20	17.59
d_w (min)	Grade A	5.9	6.9	8.9	11.6	14.6
	Grade B	—	6.7	8.7	11.4	14.4
c (max)		0.4	0.5	0.5	0.6	0.6
a (max)		2.1	2.4	3	3.75	4.5
r (max)		0.2	0.2	0.25	0.4	0.4
$\dfrac{l}{b}$ GB 5782—2000		$\dfrac{25\sim40}{14}$	$\dfrac{25\sim50}{16}$	$\dfrac{30\sim60}{18}$	$\dfrac{35\sim80}{18}$	$\dfrac{40\sim100}{26}$
$\dfrac{l}{b}$ GB 5783—2000		$\dfrac{8\sim40}{l\sim a}$	$\dfrac{10\sim50}{l\sim a}$	$\dfrac{12\sim60}{l\sim a}$	$\dfrac{16\sim80}{l\sim a}$	$\dfrac{20\sim100}{l\sim a}$
d_1 GB 31.1—88	max	—	—	1.85	2.25	2.75
	min	—	—	1.6	2	2.5

Appendix 2

(Continued)

Nominal diam. d		M4	M5	M6	M8	M10
$\dfrac{l}{l_k}$ GB 31.1—2000		—	—	$\dfrac{30 \sim 60}{27 \sim 57}$	$\dfrac{35 \sim 80}{31 \sim 76}$	$\dfrac{40 \sim 100}{36 \sim 96}$
s		18	24	30	36	46
k		7.5	10	12.5	15	18.7
$e(\min)$	Grade A	20.03	26.75	33.53	39.98	—
	Grade B	19.85	26.17	32.95	39.55	50.85
$d_w(\min)$	Grade A	16.6	22.5	28.2	33.6	—
	Grade B	16.4	22	27.7	33.2	—
$c(\max)$		0.6	0.8	0.8	0.8	0.8
$a(\max)$		2.25	6	7.5	9	10.5
$r(\max)$		0.6	0.6	0.8	0.8	1
$\dfrac{l}{b}$ GB 5782—2000		$\dfrac{45 \sim 120}{30}$	$\dfrac{55 \sim 120}{38}$ $\dfrac{130 \sim 160}{44}$	$\dfrac{65 \sim 120}{46}$ $\dfrac{130 \sim 200}{52}$	$\dfrac{80 \sim 120}{54}$ $\dfrac{130 \sim 200}{60}$ $\dfrac{220 \sim 240}{52}$	$\dfrac{90 \sim 120}{66}$ $\dfrac{130 \sim 200}{72}$ s
$\dfrac{l}{b}$ GB 5783—86		$\dfrac{25 \sim 100}{l \sim a}$	$\dfrac{35 \sim 100}{l \sim a}$	$\dfrac{40 \sim 100}{l \sim a}$	$\dfrac{40 \sim 100}{l \sim a}$	$\dfrac{40 \sim 100}{l \sim a}$
d_1 GB 31.1 —88	max	3.5	4.3	4.3	5.3	6.6
	min	3.2	4	4	5	6.3
$\dfrac{l}{l_h}$ GB 31.1—2000		$\dfrac{45 \sim 120}{40 \sim 115}$	$\dfrac{55 \sim 160}{49 \sim 154}$	$\dfrac{65 \sim 200}{59 \sim 194}$	$\dfrac{80 \sim 240}{73 \sim 233}$	$\dfrac{90 \sim 300}{81 \sim 291}$
L series		6,8,10,12,16,20,25,30,35,40,45,50,(55),60,(65),70,80,90,100,110,120,130,140,150,160,180,200,220,240,260,280,300				

3.3 Nuts

3.3.1 Type I hex. nuts—grade c (GB/T 41—2000)

Lable Examples

Nominal diam D = M12, grade C, Type I hex. Nut: Nut GB/T 41　M12

249

Fundamentals of Engineering Drawing

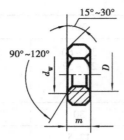
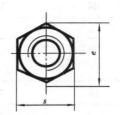

Fig. 11

Table 14 Type I hex. nuts (GB/T 41—2000)   (mm)

Nominal diam. D	M5	M6	M8	M10	M12	M16	M20	M24	M30
d_w(min)	6.9	8.7	11.5	14.5	16.5	22	27.7	33.2	42.7
e(min)	8.63	10.79	14.20	17.59	19.85	26.17	32.95	39.55	50.85
m(max)	5.6	6.1	7.9	9.5	12.2	15.9	18.7	22.3	26.4
s(max)	8	10	13	16	18	24	30	36	46

3.3.2 Type I hex. nuts—grade A and grade B (GB/T 6170—2000)

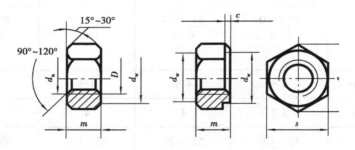

Fig. 12

Lable Examples

Nut : Nominal diam D = M12, grade A and type I

Nut GB/T 6170   M12

Nut : Nominal diam D = M12, grade B and type I

Nut GB/T6170   M12

Table 15 Type I hex. nuts (GB/T 6170—2000)   (mm)

Nominal diam. D	C	d_a(min)	d_w(min)	m(max)	s(max)	e(min)
M4	0.4	4	5.9	3.2	7	7.66
M5	0.5	5	6.9	4.7	8	8.79
M6	0.5	6	8.9	5.2	10	11.05
M8	0.6	8	11.6	6.8	13	14.38

(Continued)

Nominal diam. D	C	d_a(min)	d_w(min)	m(max)	s(max)	e(min)
M10	0.6	10	14.6	8.4	16	17.77
M12	0.6	12	16.6	10.8	18	20.03
M16	0.8	16	22.5	14.8	24	32.95
M20	0.8	20	27.7	18	30	32.95
M24	0.8	24	33.2	21.5	36	39.55
M30	0.8	30	42.7	25.6	46	50.85

3.3.3 Type I Hex. nuts—grade A and grade B—chamfer (GB/T 6172—2000)

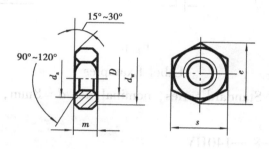

Fig. 13

Label Example

Nut : Nominal diam $D = 12$, hex.

Nut GB/T 6172 M12

Table 16 Hex. nuts (GB/T 6172—2000) (mm)

Nominal diam. D	d_a(min)	d_w(min)	m(max)	s(max)	e(min)
M4	4	5.9	2.2	7	7.66
M5	5	6.9	2.7	8	8.79
M6	6	8.9	3.2	10	11.05
M8	8	11.6	4	13	14.28
M10	10	14.6	5	16	17.77
M12	12	16.6	6	18	20.03
M16	16	22.5	8	24	26.75
M20	20	27.7	10	30	32.55
M24	24	33.2	12	36	39.55
M30	30	42.7	15	46	50.85

3.4 Washers

3.4.1 Plain washers—grade A (GB/T 97.1—1985)

3.4.2 Plain washers—grade A (GB/T 97.2—1985)

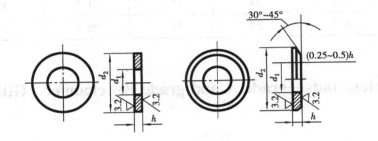

Fig. 14

Label Examples

Grade A plain washer: Standard series, nominal size $d = 8$mm, mechanical property grade: 140HV

Washer GB/T 97.1—8—140HV

Grade A plain washer: Standard series, nominal size $d = 8$mm, mechanical property grade: A140

Washer GB/T 97.1—8—A140

Grade A plain washer: Standard series, nominal size $d = 8$mm, mechanical property grade: 140HV, chamfer

Washer GB/T 97.2—8—140HV

Table 17 Plain washers (GB/T 97.1—1985), (GB/T 97.1—1985) (mm)

Nominal size (Nominal diam. d)	Inside diam. d_1	Outside diam. d_2	Thickness h
4	4.3	9	0.8
5	5.3	10	1
6	6.4	12	1.6
8	8.4	16	1.6
10	10.5	20	2
12	13	24	2.5
14	15	28	2.5
16	17	30	3
20	21	37	3
24	25	44	4
30	31	56	4

3.4.3 Plain washers—grade C (GB/T 95—1985)

Label Examples

Grade C plain washer: Standard, Nominal size $d=8$mm, mechanical property grade: 100HV,

Washer GB/T 95—8—100HV

Fig. 15

Table 18 Plain washers (GB/T 95—1985)   (mm)

Nominal size (Nominal diam. d)	Inside diam. d_1	Outside diam. d_2	Thickness h
5	5.5	10	1
6	6.6	12	1.6
8	9	16	1.6
10	11	20	2
12	13.5	24	2.5
14	15.5	28	2.5
16	17.5	30	3
20	22	37	3
24	26	44	4
30	33	56	4

3.4.4 Helix spring Lock washers (GB/T 93—1987)

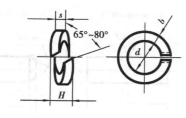
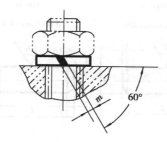

Fig. 16

Label Examples

Standard spring lock Washer (GB/T 859—1987): The nominal diam. of thread: 16mm

Washer GB/T 859—16

Light helix spring lock washer (GB/T 93—1987): The nominal diam.: 12mm

Washer GB/T 93—16

Table 19 Helix spring lock washers (GB/T 93—1987) (mm)

Nominal size (Nominal diam.)	d	H		s,b	s	m		b
		GB 93 —87	GB 859 —87	GB 93 —87	GB 859 —87	GB 93 —87	GB 859 —87	GB 859 —87
4	4.1	2.2	1.6	1.1	0.8	0.55	0.4	1.2
5	5.1	2.6	2.2	1.3	1.1	0.65	0.55	1.5
6	6.1	3.2	2.6	1.6	1.3	0.8	0.65	2
8	8.1	4.2	3.2	2.1	1.6	1.05	0.8	2.5
10	10.2	5.2	4	2.6	2	1.3	1	3
12	12.2	6.2	5	3.1	2.5	1.55	1.25	3.5
(14)	14.2	7.2	6	3.6	3	1.8	1.5	4
16	16.2	8.2	6.4	4.1	3.2	2.05	1.6	4.5
(18)	18.2	9	7.2	4.5	3.6	2.25	1.8	5
20	20.2	10	8	5	4	2.5	2	5.5
(22)	22.5	11	9	5.5	4.5	2.75	2.25	6
24	24.5	12	10	6	5	3	2.5	7
(27)	27.5	13.6	11	6.8	5.5	3.4	2.75	8
30	30.5	15	12	7.5	6	3.75	3	9

3.5 Pins

3.5.1 Cylinder pins (GB/T 119—1986)

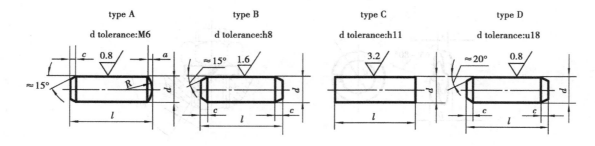

Fig. 17

Label Examples

Type A straight Pin: Nominal diam, $d=8$mm, Length $L=30$mm

Pin GB/T 119　A8 ×30

Type B straight Pin: Nominal diam $d=10$mm Length $L=40$mm

Pin GB/T 119   B10 ×40

Table 20   Cylinder pins (GB/T 119—1986)   (mm)

d	1	1.2	1.5	2	2.5	3	4	5
$a\approx$	0.12	0.16	0.20	0.25	0.30	0.40	0.50	0.63
$c\approx$	0.20	0.25	0.30	0.35	0.40	0.50	0.63	0.80
l ranges	4~10	4~12	4~16	6~20	6~24	8~30	8~40	10~15
d	6	8	10	12	16	20	25	30
$a\approx$	0.80	1.0	1.2	1.6	2.0	2.5	3.0	4.0
$c\approx$	1.2	1.6	2.0	2.5	3.0	3.5	4.0	5.0
l ranges	12~60	14~80	18~95	22~140	26~180	35~200	50~200	60~200

3.5.2 Taper pins (GB/T 117—1986)

Type A   Type B

$$R_1 \approx d$$
$$R_2 \approx d + \frac{l-2a}{50}$$

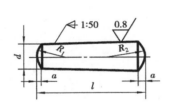
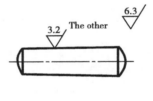

Fig. 18

Label Example

Type A taper Pin: Nominal size $d = 10$mm, length $L = 60$mm,

Pin GB/T 117 A10 ×60

Table 21   Taper pins (GB/T 117—1986)   (mm)

d	1	1.2	1.5	2	2.5	3	4	5
$a\approx$	0.12	0.16	0.2	0.25	0.3	0.4	0.5	0.63
l ranges	6~16	6~20	8~24	10~35	10~35	12~35	14~55	10~60
d	6	8	10	12	16	20	25	30
$a\approx$	0.8	1	1.2	1.6	2	2.5	3	4
l ranges	22~90	22~120	26~160	32~180	40~200	45~200	50~200	55~200

3.5.3 Cotter pins (GB/T 91—1986)

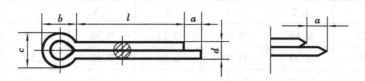

Fig. 19

Label Example

Nominal size $d = 5$mm, length $L = 50$mm: Pin GB/T 91—1986 5×50

Table 22 Cotter pins (GB/T 91—1986)   (mm)

d	2	2.5	3.2	4	5	6.3	8	10	12
e	3.6	4.6	5.8	7.4	9.2	11.8	15	19	24.8
$b \approx$	4	5	6.4	8	10	12.6	16	20	26
a	2.5	2.5	3.2	4	4	4	4	6.3	6.3
l ranges	10~40	12~50	14~65	18~80	22~100	30~120	40~160	45~200	70~200
l series	10,12,14,16,20,22,26,28,30,32,36,40,45,50,55,60,65,70,75,80,85,90,95,100,120,140,160,180,200								

Note: The norminal diameter of pin hole is d.

3.6 Keys

3.6.1 Flat keys

3.6.1.1 Dimensions of cross-section of keys, keyseats and keyways (GB/T 1095—1997)

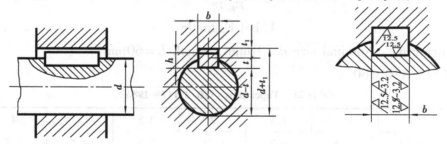

Fig. 20

Note: in engineering drawings, the height of keyseat (轴槽) is $(d-t)$ while the height of keyway (轮毂槽) is $(d+t_1)$.

Appendix 2

Table 23 Dimensions of cross-section of keys, keyseats and keyways (GB/T 1095—1997) (mm)

Shafts	Keys	Keyseats and keyways								
Norminal Diam. d	Norminal Dim. $B \times l$	Norminal Dim. b	Depth b					Heights		
			Limit deviation					Shaft t	Hub $t1$	
			Looser key joining		General key joining		Tight key joining			
			Shaft H9	Hub D10	Shaft N9	Hub Js9	Shaft, hub P9	Nor. Diam.	Limit deviation	Nor. Diam.	Limit deviation
From 6~8	2×2	2	+0.025	+0.060	−0.004		−0.006	1.2	+0.1	1	+0.1
>8~10	3×3	3	0	+0.020	−0.029	±0.0125	−0.031	1.8		1.4	
>10~12	4×4	4	+0.030	+0.078	0		−0.012	2.5	0	1.8	0
>12~17	5×5	5				±0.015		3.0		2.3	
>17~22	6×6	6	0	+0.030	−0.030		−0.042	3.5		2.8	
>22~30	8×7	8	+0.036	+0.098	0	±0.018	−0.015	4.0	+0.2	3.3	+0.2
>30~38	10×8	10	0	+0.040	−0.036		−0.051	5.0		3.3	
>38~44	12×8	12						5.0		3.3	
>44~50	14×9	14	+0.043	+0.120	0	±0.0215	−0.018	5.5	0	3.8	0
>50~58	16×10	16	0	+0.050	−0.043		−0.061	6.0		4.3	
>58~65	18×11	18						7.0		4.4	
>65~75	20×12	20						7.5		4.9	+0.2
>75~85	22×14	22	+0.052	+0.149	0	±0.026	−0.022	9.0	+0.2	5.4	
>85~95	25×14	25	0	+0.065	−0.052		−0.074	9.0	0	5.4	
>95~110	28×16	28						10.0		6.4	0
>110~130	32×18	32						11.0		7.4	
>130~150	36×20	36	+0.062	+0.180	0	±0.031	−0.026	12.0	+0.3	8.4	+0.3
>150~170	40×22	40	0	+0.080	+0.062		−0.088	13.0	0	9.4	
>170~200	45×25	45						15.0		10.4	0

3.6.1.2 Size dimension (GB/T 1096—1979)

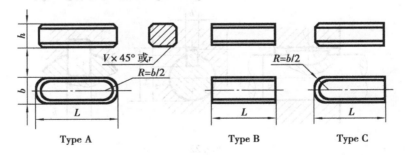

Fig. 21

Label Example

Flat key (Type A): $b = 16$mm, $h = 10$mm, $l = 100$mm,

Key 16×100 GB/T 1096—1979

Flat key (Type B): $b = 16$mm, $h = 10$mm, $l = 100$mm,

Key C 16×100 GB/T 1096—1979

Table 24 Size dimensions (GB/T 1096—1979)   (mm)

	Norminal sizes	2	3	4	5	6	8	10	12	14	16	18	20	22
b	Limit deviations h9	0 −0.025		0 −0.030			0 −0.036			0 −0.043			0 −0.052	
h	Norminal sizes	2	3	4	5	6	7	8	8	9	10	11	12	14
	Limit deviations h11	0 −0.06 (0 −0.025)		0 −0.075 (0 −0.030)			0 −0.090				0 −0.110			
	c or r	0.16~0.25		0.25~0.40			0.40~0.60				0.60~0.80			
	L	6~20	6~36	8~45	10~56	14~70	18~90	22~110	28~140	36~160	45~180	50~200	56~220	63~250

Length series and tolerance

	Norminal sizes	6	8	10	12	14	16	18	20	22	25	28	32
L	Limit deviations h14	0 −0.30					0 −0.43				0 −0.52		0 −0.62
	Norminal sizes	36	40	45	50	56	63	70	80	90	100	110	125
L	Limit deviations h14	0 −0.62					0 −0.74			0 −0.87			0 −1.0
	Norminal sizes	140	160	180	200	220	250	280	320	360	400	450	500
L	Limit deviations h14	0 −1.0					0 −1.15			0 −130			0 −1.55

3.6.2 Woodruff keys

3.6.2.1 Dimensions of cross-section of keys, keyseats and keyways (GB/T 1095—1997)

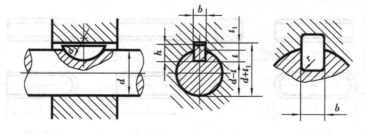

Fig. 22

Note: In engineering drawings, the height of keyseat (轴槽) is $(d - t)$ while the height of keyway (轮毂槽) is $(d + t_1)$.

Table 25 Dimensions of cross-section of keys, keyseats and keyways (GB/T 1098—1997)   (mm)

Shafts	Keys	Keyseats and keyways								
Norminal Diam. d	Norminal Dim. $B \times l$	Norminal Dim. b	Depth b				Heights		Hub $t1$	
			Limit deviation				Shaft t			
			Looser key joining		General key joining		Tight key joining			
			Shaft H9	Hub D10	Shaft N9	Hub Js9	Shaft, hub P9		Nor. Diam.	Limit deviation	Nor. Diam.	Limit deviation
from 3~4	from 3~4	1.0×1.4×4	1.0	−0.004 −0.029	±0.012	−0.006 −0.031	1.0	+0.10	0.6	0.08	0.16
>4~5	>4~6	1.5×2.6×7	1.5				2.0		0.8		
>5~6	>6~8	2.0×2.6×7	2.0				1.8		1.0		
>6~7	>8~10	2.0×3.7×10	2.0				2.9		1.0		
>7~8	>10~12	2.5×3.7×10	2.5				2.7		1.2		
>8~10	>12~15	3.0×5.0×13	3.0				3.8	+0.10	1.4		
>10~12	>15~18	3.0×6.5×16	3.0				5.3		1.4		
>12~14	>18~20	4.0×6.5×16	4.0	0 −0.030	±0.015	−0.012 −0.042	5.0	+0.20	1.8	0.16	0.25
>14~16	>20~22	4.0×7.5×19	4.0				6.0		1.8		
>16~18	>22~25	5.0×6.5×16	5.0				4.5		2.3		
>18~20	>25~28	5.0×7.5×19	5.0				5.5		2.3		
>20~22	>28~32	5.0×8.0×22	5.0				6.5		2.3		
>22~25	>32~36	6.0×9.0×22	6.0				7.0		2.8		
>25~28	>36~40	6.0×10.0×25	6.0				7.5	+0.30	2.8		
>28~32	40	8.0×11.0×28	8.0	0 −0.036	±0.018	−0.015 −0.051	8.0		3.3	0.25	0.40
>32~38	—	10.0×13.0×22	10.0				10.0		3.3		

3.6.2.2 Size dimansions (GB 1099—1979)

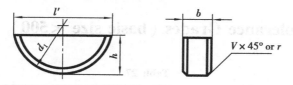

Fig. 23

Label Example

Woodruff key : $b = 16$mm, $h = 10$mm, $d_1 = 25$mm,
Key 6 × 25   GB/T 1099—1979

Fundamentals of Engineering Drawing

Table 26 Size dimansions (GB/T 1099—1979) (mm)

Depth		Height		Diameter		$l \approx$	C	
Norminal sizes	Limit deviations (h9)	Norminal sizes	Limit deviations (h11)	Norminal sizes	Limit deviations (h12)		最小	最大
1.0	0 −0.025	1.4	0 −0.060	4	0 −0.120	3.9	0.16	0.25
1.5		2.6		7		6.8		
2.0		2.6		7		6.8		
2.0		3.7	0 −0.075	10	0 −0.150	9.7		
2.5		3.7		10		9.7		
3.0		5.0		13		12.7		
3.0		6.5		16	0 −0.180	15.7		
4.0	0 −0.030	6.5		16		15.7	0.25	0.40
4.0		7.5		19	0 −0.210	18.6		
5.0		6.5	0 −0.090	16	0 −0.180	15.8		
5.0		7.5		19		18.6		
5.0		9.0		22		21.6		
6.0		9.0		22	0 −0.210	21.6		
6.0		10.0		25		24.5		
8.0	0 −0.036	11.0	0 −0.110	28		27.4	0.40	0.60
10.0		13.0		32	0 −0.250	31.4		

4. Limits and Fits

4.1 International Tolerance Grades (basic size <500)

Table 27

Basic sizes mm		Tolerance grades									
		IT01	IT0	IT1	IT2	IT3	IT4	IT5	IT6	IT7	IT8
From	To	(μm)									
—	3	0.3	0.5	0.8	1.2	2	3	4	6	10	14
3	6	0.4	0.6	1	1.5	2.5	4	5	8	12	18
6	10	0.4	0.6	1	1.5	2.5	4	6	9	15	22
10	18	0.5	0.8	1.2	2	3	5	8	11	18	27
18	30	0.6	1	1.5	2.5	4	6	9	13	21	33
30	50	0.6	1	1.5	2.5	4	7	11	16	25	39

(Continued)

Basic sizes mm		Tolerance grades									
From	To	IT01	IT0	IT1	IT2	IT3	IT4	IT5	IT6	IT7	IT8
						(μm)					
50	80	0.8	1.2	2	3	5	8	13	19	30	46
80	120	1	1.5	2.5	4	6	10	15	22	35	54
120	180	1.2	2	3.5	5	8	12	18	25	40	63
180	250	2	3	4.5	7	10	14	20	29	46	72
250	315	2.5	4	6	8	12	16	23	32	52	81
315	400	3	5	7	9	13	18	25	36	57	89
400	500	4	6	8	10	15	20	27	40	63	97

Basic sizes mm		Tolerance grades									
From	To	IT9	IT10	IT11	IT12	IT13	IT14	IT15	IT16	IT17	IT18
		(μm)			(mm)						
—	3	25	40	60	0.10	0.14	0.25	0.40	0.60	1.0	1.4
3	6	30	48	75	0.12	0.18	0.30	0.48	0.75	1.2	1.8
6	10	36	58	90	0.15	0.22	0.36	0.58	0.90	1.5	2.2
10	18	43	70	110	0.18	0.27	0.43	0.70	1.10	1.8	2.7
18	30	52	84	130	0.21	0.33	0.52	0.84	1.20	2.0	3.3
30	50	62	100	160	0.25	0.39	0.62	1.00	1.60	2.5	3.9
50	80	74	120	190	0.30	0.46	0.74	1.20	1.90	3.0	4.6
80	120	87	140	220	0.35	0.54	0.87	1.40	2.20	3.5	5.4
120	180	100	160	250	0.40	0.63	1.00	1.60	2.50	4.0	6.3
180	250	115	185	290	0.46	0.72	1.15	1.85	2.90	4.6	7.2
250	315	130	210	320	0.52	0.81	1.30	2.10	3.20	5.2	8.1
315	400	140	230	360	0.57	0.89	1.40	2.30	3.60	5.7	8.9
400	500	155	250	400	0.63	0.97	01.55	2.50	4.00	6.3	9.7

4.2 Preferential and Common Fits in Basic-Hole

Table 28

Basic-hole	Shafts																		
	a	b	c	d	e	f	g	h	js	k	m	n	p	r	s	t	u	v	x	y	z
	Clearance fits								Transition fits			Interference fits								
H6						$\frac{H6}{f5}$	$\frac{H6}{g5}$	$\frac{H6}{h5}$	$\frac{H6}{js5}$	$\frac{H6}{k5}$	$\frac{H6}{m5}$	$\frac{H6}{n5}$	$\frac{H6}{p5}$	$\frac{H6}{r5}$	$\frac{H6}{s5}$	$\frac{H6}{t5}$				
H7						$\frac{H7}{f6}$	●$\frac{H7}{g5}$	●$\frac{H7}{h6}$	$\frac{H7}{js6}$	●$\frac{H7}{k6}$	$\frac{H7}{m6}$	●$\frac{H7}{n6}$	●$\frac{H7}{p6}$	$\frac{H7}{r6}$	●$\frac{H7}{s6}$	$\frac{H7}{t6}$	●$\frac{H7}{u6}$	$\frac{H7}{v6}$	$\frac{H7}{x6}$	$\frac{H7}{y6}$	$\frac{H7}{z6}$

(Continued)

Basic-hole	Shafts																				
	a	b	c	d	e	f	g	h	js	k	m	n	p	r	s	t	u	v	x	y	z
	Clearance fits								Transition fits				Interference fits								
H8					$\frac{H8}{e7}$	● $\frac{H8}{f7}$	$\frac{H8}{g7}$	● $\frac{H8}{H7}$	$\frac{H8}{js7}$	$\frac{H8}{k7}$	$\frac{H8}{m7}$	$\frac{H8}{n7}$	$\frac{H8}{p7}$	$\frac{H8}{r7}$	$\frac{H8}{s7}$	$\frac{H8}{t7}$	$\frac{H8}{u7}$				
				$\frac{H8}{d8}$	$\frac{H8}{e8}$	$\frac{H8}{f8}$		$\frac{H8}{H8}$													
H9			$\frac{H9}{c9}$	● $\frac{H9}{d9}$	$\frac{H9}{e9}$	$\frac{H9}{f9}$		● $\frac{H9}{H9}$													
H10			$\frac{H10}{c10}$	$\frac{H10}{d10}$				$\frac{H10}{H10}$													
H11	$\frac{H11}{a10}$	$\frac{H11}{b10}$	● $\frac{H11}{c10}$	$\frac{H11}{d10}$				● $\frac{H11}{H10}$													
H12		$\frac{H12}{b12}$						$\frac{H12}{h12}$													

4.3 Preferential and Common Fits in Basic-Shaft

Table 29

Basic-shaft	Holes																				
	A	B	C	D	E	F	G	H	Js	K	M	N	P	R	S	T	U	V	X	Y	Z
	Clearance fits								Transition fits				Interference fits								
h5						$\frac{F6}{h5}$	$\frac{G6}{h5}$	$\frac{H6}{h5}$	$\frac{JS6}{h5}$	$\frac{K6}{h5}$	$\frac{M6}{h5}$	$\frac{N6}{h5}$	$\frac{P6}{h5}$	$\frac{R6}{h5}$	$\frac{S6}{h5}$	$\frac{T6}{h5}$					
h6						$\frac{F7}{h6}$	● $\frac{G7}{h6}$	● $\frac{H7}{h6}$	$\frac{JS7}{h6}$	● $\frac{K7}{h6}$	$\frac{M7}{h6}$	● $\frac{N7}{h6}$	● $\frac{P7}{h6}$	$\frac{R7}{h6}$	● $\frac{S7}{h6}$	$\frac{T7}{h6}$	● $\frac{U7}{h6}$				
h7					$\frac{E8}{h7}$	● $\frac{F8}{h7}$		● $\frac{H8}{h7}$	$\frac{JS8}{h7}$	$\frac{K8}{h7}$	$\frac{M8}{h7}$	$\frac{N8}{h7}$									
h8				$\frac{D8}{h8}$	$\frac{E8}{h8}$	$\frac{F8}{h8}$		$\frac{H8}{h8}$													
h9				● $\frac{D9}{h9}$	$\frac{E9}{h9}$	$\frac{F9}{h9}$		● $\frac{H9}{h9}$													
h10				$\frac{D10}{h10}$				$\frac{H10}{h10}$													
h11	$\frac{A11}{h11}$	$\frac{B11}{h11}$	● $\frac{C11}{h11}$	$\frac{D11}{h11}$				● $\frac{H11}{h11}$													
h12		$\frac{B12}{h12}$						$\frac{H12}{h12}$													

4.4 Limits Deviations of Shafts

Table 30 (μm)

Fundamental deviations		*a*	*b*		*c*			*d*			*e*			*f*		
Tolerance grades		11	11	12	9	10	11	9	10	11	7	8	9	5	6	7
Basic sizes – mm		Tolerance zone														
From	To															
—	3	−270 −330	−140 −200	−140 −240	−60 −85	−60 −120	−60 −120	−20 −45	−20 −60	−20 −80	−14 −24	−14 −28	−14 −39	−6 −10	−6 −10	−6 −16
3	6	−270 −345	−140 −215	−140 −260	−70 −100	−70 −118	−70 −145	−30 −60	−30 −78	−30 −105	−20 −32	−20 −38	−20 −50	−10 −15	−10 −18	−10 −22
6	10	−280 −370	−150 −240	−150 −300	−80 −116	−80 −138	−80 −170	−40 −76	−40 −98	−40 −130	−25 −40	−25 −47	−25 −61	−13 −19	−13 −22	−13 −28
10	14	−290 −400	−150 −260	−150 −330	−95 −138	−95 −165	−95 −205	−50 −93	−50 −120	−50 −160	−32 −50	−32 −59	−32 −75	−16 −24	−16 −27	−16 −34
14	18															
18	24	−300 −430	−160 −290	−160 −370	−110 −162	−110 −194	−110 −240	−65 −117	−65 −149	−65 −195	−40 −61	−40 −73	−40 −92	−20 −29	−20 −33	−20 −41
24	30															
30	40	−310 −470	−170 −330	−170 −420	−120 −182	−120 −182	−120 −280	−80 −142	−80 −180	−80 −240	−50 −75	−50 −89	−50 −112	−25 −36	−25 −41	−25 −50
40	50	−320 −480	−190 −340	−180 −430	−130 −192	−130 −230	−130 −290									
50	65	−340 −530	−190 −380	−190 −490	−140 −214	−140 −260	−140 −330	−100 −174	−100 −220	−100 −290	−60 −90	−60 −106	−60 −134	−30 −43	−30 −49	−30 −60
65	80	−360 −550	−200 −390	−200 −500	−150 −224	−150 −270	−150 −340									
80	100	−380 −600	−220 −440	−220 −570	−170 −257	−170 −310	−170 −390	−120 −207	−120 −260	−120 −340	−72 −107	−72 −126	−72 −159	−36 −51	−36 −58	−36 −71
100	120	−410 −630	−240 −460	−240 −590	−180 −267	−180 −320	−1800 −400									
120	140	−460 −710	−260 −510	−260 −660	−200 −300	−200 −360	−200 −450	−145 −245	−150 −305	−145 −395	−85 −125	−85 −148	−85 −185	−43 −61	−43 −68	−43 −83
140	160	−520 −770	−280 −530	−280 −680	−210 −310	−210 −370	−210 −460									
160	180	−580 −830	−310 −560	−310 −710	−230 −330	−230 −390	−230 −480									
180	200	−660 −950	−340 −630	−340 −800	−240 −355	−240 −425	−240 −530	−170 −285	−170 −355	−170 −460	−100 −146	−100 −172	−100 −215	−50 −70	−50 −79	−50 −96
200	225	−740 −1030	−380 −670	−380 −840	−260 −375	−260 −445	−260 −550									
225	250	−820 −1110	−420 −710	−420 −880	−280 −395	−280 −465	−280 −570									
250	280	−920 −1240	−480 −800	−480 −1000	−300 −430	−300 −510	−300 −620	−190 −320	−190 −400	−190 −510	−110 −162	−110 −240	−110 −240	−56 −79	−56 −88	−56 −108
280	315	−1059 −1370	−540 −860	−540 −1060	−330 −460	−330 −540	−330 −650									
315	355	−1200 −1560	−680 −960	−600 −1175	−360 −500	−360 −590	−360 −720	−210 −350	−210 −440	−210 −570	−125 −182	−1235 −214	−125 −265	−62 −87	−62 −98	−62 −119
355	400	−1350 −1710	−680 −1040	−680 −1250	−400 −540	−400 −630	−400 −760									
400	450	−1500 −1900	−760 −1160	−760 −1390	−440 −595	−440 −690	−440 −840	−230 −385	−230 −480	−230 −630	−135 −198	−135 −232	−135 −290	−68 −95	−68 −108	−68 −131
450	500	−1650 −2050	840 −1240	−840 −1470	−480 −635	−480 −730	−480 −880									

(Continued)

Fundamental deviations	f		g			h								js			k
Tolerance grades	8	9	5	6	7	5	6	7	8	9	10	11	12	5	6	7	6
Basic sizes – mm From / To	Tolerance zone																
— / 3	−6 / −20	−6 / −31	−2 / −6	−2 / −8	−2 / −12	0 / −4	0 / −6	0 / −10	0 / −14	0 / −25	0 / −40	0 / −60	0 / −100	±2	±3	±5	+6 / 0
3 / 6	−10 / −28	−10 / −40	−4 / −9	−4 / −12	−4 / −16	0 / −5	0 / −8	0 / −12	0 / −18	0 / −30	0 / −48	0 / −75	0 / −120	±2.5	±4	±6	+9 / +1
6 / 10	−13 / −35	−13 / −49	−5 / −11	−5 / −14	−5 / −20	0 / −6	0 / −9	0 / −15	0 / −22	0 / −36	0 / −58	0 / −90	0 / −150	±3	±4.5	±7	+120 / +1
10 / 14, 14 / 18	−16 / −43	−16 / −59	−6 / −14	−6 / −17	−6 / −24	0 / −8	0 / −11	0 / −18	0 / −27	0 / −43	0 / −70	0 / −110	0 / −180	±4	±5.5	±9	+12 / +1
18 / 24, 24 / 30	−20 / −53	−20 / −72	−7 / −16	−7 / −20	−7 / −28	0 / −9	0 / −13	0 / −21	0 / −33	0 / −52	0 / −84	0 / −130	0 / −210	±4.5	±6.5	±10	+15 / +2
30 / 40, 40 / 50	−25 / −64	−25 / −87	−9 / −20	−9 / −25	−9 / −34	0 / −11	0 / −16	0 / −25	0 / −39	0 / −62	0 / −100	0 / −160	0 / −250	±5.5	±8	±12	+18 / +2
50 / 65, 65 / 80	−30 / −76	−30 / −104	−10 / −23	−10 / −29	−10 / −40	0 / −13	0 / −19	30 / −46	0 / −74	0 / −120	0 / −190	0 / −300		±6.5	±9.5	±15	+21 / +2
80 / 100, 100 / 120	−36 / −90	−36 / −123	−12 / −27	−12 / −34	−12 / −47	0 / −15	0 / −22	0 / −35	0 / −54	0 / −87	0 / −140	0 / −220	0 / −350	±7.5	±11	±17	+25 / +3
120 / 140, 140 / 160, 160 / 180	−43 / −106	−43 / −143	−14 / −32	−14 / −39	−14 / −54	0 / −18	0 / −25	0 / −40	0 / −63	0 / −100	0 / −160	0 / −250	0 / −400	±9	±12.5	±20	+28 / +3
180 / 200, 200 / 225, 225 / 250	−50 / −122	−50 / −165	−15 / −35	15 / −44	−15 / −61	0 / −20	0 / −29	0 / −46	0 / −72	0 / −115	0 / −185	0 / −290	0 / −460	±10	±14.5	±23	+33 / +4
250 / 280, 280 / 315	−56 / −137	−56 / −186	−17 / −40	−17 / −49	−17 / −69	0 / −23	0 / −32	0 / −52	0 / −81	0 / −130	0 / −210	0 / −320	0 / −520	±11.5	±16	±26	+36 / +4
315 / 355, 355 / 400	−62 / −151	−62 / −202	−18 / −43	−18 / −54	−18 / −75	0 / −25	0 / −36	0 / −57	0 / −89	0 / −140	0 / −230	0 / −360	0 / −570	±12.5	±18	±28	+40 / +4
400 / 450, 450 / 500	−68 / −165	−68 / −223	−20 / −47	−20 / −60	−20 / −83	0 / −27	0 / −40	0 / −63	0 / −97	0 / −155	0 / −250	0 / −400	0 / −630	±13.5	±20	±31	+45 / +5

264

Appendix 2

Fundamental deviations		k	m			n			p			r			s
Tolerance grades		7	5	6	7	5	6	7	5	6	7	5	6	7	5
Basic sizes – mm		Tolerance zone													
From	To														
0	3	+10 0	+6 +2	+8 +2	+12 +2	+8 +4	+10 +4	+14 +4	+10 +6	+12 +6	+16 +6	+14 +10	+16 +10	+20 +10	+18 +14
3	6	+13 +1	+9 +4	+12 +4	+16 +4	+13 +8	+16 +8	+20 +8	+17 +12	+20 +12	+24 +12	+20 +15	+23 +15	+27 +15	+24 +19
6	10	+16 +1	+12 +6	+15 +6	+21 +6	+16 +10	+19 +10	+25 +10	+21 +15	+24 +15	+30 +15	+25 +19	+28 +19	+34 +19	+29 +23
10	14	+19 +1	+15 +7	+18 +7	+25 +7	+20 +12	+23 +12	+30 +12	+26 +18	+29 +18	+36 +18	+31 +23	+34 +23	+41 +23	+36 +28
14	18														
18	24	+23 +2	+17 +8	+21 +8	+29 +8	+24 +15	+28 +15	+36 +15	+31 +22	+35 +22	+43 +22	+37 +28	+41 +28	+49 +28	+44 +35
24	30														
30	40	+27 +2	+20 +9	+25 +9	+34 +9	+28 +17	+33 +17	+42 +17	+37 +26	+42 +26	+51 +26	+45 +34	+50 +34	+59 +34	+54 +43
40	50														
50	65	+23 +2	+24 +11	+30 +11	+41 +11	+33 +20	+39 +20	+50 +20	+45 +32	+51 +32	+62 +32	+54 +41	+60 +41	+71 41	+66 +53
65	80											+56 +43	+62 +43	+73 +43	+72 +59
80	100	+38 +3	+28 +13	+35 +13	+48 +13	+38 +23	+45 +23	+58 +23	+52 +37	+59 +37	+72 +37	+69 +51	+73 +51	+86 51	+86 +71
100	120											+69 +54	+76 +54	+89 +54	+94 +79
120	140	+43 +3	+33 +15	+40 +15	+55 +15	+45 +27	+52 +27	+67 +27	+61 +43	+68 +43	+83 +43	+81 +63	+88 +63	+103 63	+110 +92
140	160											+83 +65	+90 +65	+105 +65	+118 +100
160	180											+86 68	+93 +68	+108 +68	+126 +108
180	200	+50 +4	+37 +17	+46 +17	+63 +17	+51 +31	+60 +31	+77 +31	+70 +50	+79 +50	+96 +50	+97 +77	+106 +77	+123 77	+142 +122
200	225											+100 +80	+109 +80	+126 +80	+150 +130
225	250											+104 +84	+113 +84	+130 +84	+160 +140
250	280	+56 +4	+43 +20	+52 +20	+72 +20	+57 +34	+66 +34	+86 +34	+79 +56	+88 +56	+108 +56	+117 +94	+126 +94	+146 94	+181 +158
280	315											+121 +98	+130 +98	+150 +98	+193 +170
315	355	+61 +4	+46 +21	+57 +21	+78 +21	+62 +37	+73 +37	+94 +37	+87 +62	+98 +62	+119 +62	+133 +108	+144 +108	+165 108	+215 +190
355	400											+139 +114	+150 +114	+171 +114	+233 +208
400	450	+68 +5	+50 +23	+63 +23	+86 +23	+67 +40	+80 +40	+103 +40	+95 +68	+108 +68	+131 +68	+153 +126	+166 +126	+189 126	+259 +232
450	500											+159 +132	+172 +132	+195 +132	+279 +252

4.5 Limits Deviations of Holes

Table 31 (μm)

Fundamental deviations	A	B	C	D				E		F				G
Tolerance grades	11	11	12	11	8	9	10	11	8	9	6	7	8	9	6
Basic sizes −mm From / To						Tolerance zone									
— / 3	+330 / +270	+200 / +140	+240 / +140	+120 / +60	+34 / +20	+45 / +20	+60 / +20	+80 / +20	+28 / +14	+39 / +14	+12 / +6	+12 / +6	+20 / +6	+31 / +6	+8 / +2
3 / 6	+345 / +270	+215 / +140	+260 / +140	+145 / +70	+48 / +30	+60 / +30	+78 / +30	+105 / +30	+38 / +20	+50 / +20	+18 / +10	+22 / +10	+28 / +10	+40 / +10	+12 / +4
6 / 10	+370 / +280	+240 / +150	+300 / +150	+170 / +80	+62 / +40	+76 / +40	+98 / +40	+130 / +40	+47 / +25	+61 / +25	+22 / +13	+28 / +13	+35 / +13	+49 / +13	+14 / +5
10 / 14	+400 / +290	+260 / +150	+330 / +150	+205 / +95	+77 / +50	+93 / +50	+120 / +50	+160 / +50	+59 / +32	+75 / +32	+27 / +16	+34 / +16	+43 / +16	+59 / +16	+17 / +6
14 / 18															
18 / 24	+430 / +300	+290 / +160	+370 / +160	+240 / +110	+98 / +65	+117 / +65	+149 / +65	+195 / +65	+73 / +40	+92 / +40	+33 / +20	+41 / +20	+53 / +20	+72 / +20	+20 / +7
24 / 30															
30 / 40	+470 / +310	+330 / +170	+420 / +170	+280 / +120	+119 / +80	+142 / +80	+180 / +80	+240 / +80	+89 / +50	+112 / +50	+41 / +25	+50 / +25	+64 / +25	+87 / +25	+25 / +9
40 / 50	+480 / +320	+340 / +180	+430 / +180	+290 / +130											
50 / 65	+530 / +340	+380 / +190	+490 / +190	+330 / +140	+146 / +100	+174 / +100	+220 / +100	+290 / +100	+106 / +60	+134 / +60	+49 / +30	+60 / +30	+76 / +30	+104 / +30	+29 / +10
65 / 80	+550 / +360	+390 / +200	+500 / +200	+340 / +150											
80 / 100	+600 / +380	+440 / +220	+570 / +220	+390 / +170	+174 / +120	+207 / +120	+260 / +120	+340 / +120	+126 / +72	+159 / +72	+58 / +36	+71 / +36	+90 / +36	+123 / +36	+34 / +12
100 / 120	+630 / +410	+460 / +240	+590 / +240	+400 / +180											
120 / 140	+710 / +460	+510 / +260	+660 / +260	+450 / +200	+208 / +145	+245 / +145	+305 / +145	+395 / +145	+148 / +85	+185 / +85	+68 / +43	+83 / +43	+106 / +43	+143 / +43	+39 / +14
140 / 160	+770 / +520	+530 / +280	+680 / +280	+460 / +210											
160 / 180	+830 / +580	+560 / +310	+710 / +310	+480 / +230											
180 / 200	+950 / +660	+630 / +340	+800 / +340	+530 / +240	+242 / +170	+285 / +170	+355 / +170	+460 / +170	+172 / +100	+215 / +100	+79 / +50	+96 / +50	+122 / +50	+165 / +50	+44 / +15
200 / 225	+1030 / +740	+670 / +380	+840 / +380	+550 / +260											
225 / 250	+1110 / +820	+710 / +420	+880 / +420	+570 / +280											
250 / 280	+1240 / +920	+800 / +480	+1000 / +480	+620 / +300	+271 / +190	+320 / +190	+400 / +190	+510 / +190	+191 / +110	+240 / +110	+88 / +56	+108 / +56	+137 / +56	+186 / +56	+49 / +17
280 / 315	+1370 / +1050	+860 / +540	+1060 / +540	+650 / +330											
315 / 355	+1560 / +1200	+960 / +600	+1170 / +600	+720 / +360	+299 / +210	+350 / +210	+440 / +210	+570 / +210	+214 / +125	+265 / +125	+98 / +62	+119 / +62	+151 / +612	+202 / +62	+54 / +18
355 / 400	+1710 / +1350	+1040 / +680	+1250 / +680	+760 / +400											
400 / 450	+1900 / +1500	+1160 / +760	+1390 / +760	+840 / +400	+327 / +230	+385 / +230	+480 / +230	+630 / +230	+232 / +135	+290 / +135	+108 / +68	+131 / +68	+165 / +68	+223 / +68	+60 / +20
450 / 500	+2050 / +1650	+1240 / +840	+1470 / +840	+880 / +480											

Appendix 2

Fundamental deviations		G	H							J			JS			K		
Tolerance grades		7	6	7	8	9	10	11	12	6	7	8	6	7	8	6	7	8
Basic sizes – mm		Tolerance zone																
From	To																	
—	3	+12 / +2	+6 / 0	+10 / 0	+14 / 0	+25 / 0	+40 / 0	+60 / 0	+100 / 0	+2 / −4	+4 / −6	+6 / −8	±3	±5	±7	0 / −6	0 / −10	0 / −14
3	6	+16 / +4	+8 / 0	+12 / 0	+18 / 0	+30 / 0	+48 / 0	+75 / 0	+120 / 0	+5 / −3	—	+10 / −8	±4	±6	±9	+2 / −6	+3 / −9	+5 / −13
6	10	+20 / +5	+9 / 0	+15 / 0	+22 / 0	+36 / 0	+58 / 0	+90 / 0	+150 / 0	+5 / −4	+8 / −7	+12 / −10	±4.5	±7	±11	+2 / −7	+5 / −10	+6 / −16
10	14	+24 / +6	+11 / 0	+18 / 0	+27 / 0	+43 / 0	+70 / 0	+110 / 0	+180 / 0	+6 / −5	+10 / −8	+15 / −12	±5.5	±9	±13	+2 / −9	+6 / −12	+8 / −19
14	18																	
18	24	+28 / +7	+13 / 0	+21 / 0	+33 / 0	+52 / 0	+84 / 0	+130 / 0	+210 / 0	+8 / −5	+12 / −9	+20 / −13	±6.5	±10	±16	+2 / −11	+6 / −15	+10 / −23
24	30																	
30	40	+34 / +9	+16 / 0	+25 / 0	+39 / 0	+62 / 0	+100 / 0	+160 / 0	+250 / 0	+10 / −6	+14 / −11	+24 / −15	±8	±12	±19	+3 / −13	+7 / −18	+12 / −27
40	50																	
50	65	+40 / +10	+19 / 0	+30 / 0	+46 / 0	+74 / 0	+120 / 0	+190 / 0	+300 / 0	+13 / −6	+18 / −12	+28 / −18	±9.5	±15	±23	+4 / −15	+9 / −21	+14 / −32
65	80																	
80	100	+47 / +12	+22 / 0	+35 / 0	+54 / 0	+87 / 0	+140 / 0	+220 / 0	+310 / 0	+16 / −6	+22 / −13	+34 / −20	±11	±17	±27	+4 / −18	+10 / −25	+16 / −38
100	120																	
120	140	+54 / +14	+25 / 0	+40 / 0	+63 / 0	+100 / 0	+160 / 0	+250 / 0	+400 / 0	+18 / −7	+26 / −14	+41 / −22	±12.5	±20	±31	+4 / −21	+12 / −18	+20 / −43
140	160																	
160	180																	
180	200	+61 / +15	+29 / 0	+46 / 0	+72 / 0	+115 / 0	+185 / 0	+290 / 0	+460 / 0	+22 / −7	+30 / −16	+47 / −25	±14.5	±23	±36	+5 / −24	+13 / −33	+22 / −50
200	225																	
225	250																	
250	280	+69 / +17	+32 / 0	+52 / 0	+81 / 0	+130 / 0	+210 / 0	+320 / 0	+520 / 0	+25 / −7	+36 / −16	+55 / −26	±16	±26	±40	+5 / −36	+16 / −36	+25 / −56
280	315																	
315	355	+75 / 18	+36 / 0	+57 / 0	+89 / 0	+140 / 0	+230 / 0	+360 / 0	+570 / 0	+39 / −7	+39 / −18	+60 / −29	±18	±28	±44	+7 / −29	+17 / −40	+28 / −61
355	400																	
400	450	+83 / +20	+40 / 0	+63 / 0	+97 / 0	+155 / 0	+250 / 0	+400 / 0	+630 / 0	+33 / −7	+43 / −20	+66 / −31	±20	±31	±48	+8 / −32	+18 / −45	+29 / −68
450	500																	

Fundamentals of Engineering Drawing

Fundamental deviations		M			N			P		R			S		T		U
Tolerance grades		6	7	8	6	7	8	6	7	6	7	8	6	7	6	7	7
Basic sizes − mm		Tolerance zone															
From	To																
—	3	−2 / −8	−2 / −12	+2 / −16	−4 / −10	−4 / −14	−4 / −18	−6 / −12	−6 / −16	−10 / −16	−10 / −20	−10 / −24	−14 / −20	−14 / −24	—	—	−18 / −28
3	6	−1 / −9	0 / −12	+2 / −16	−5 / −13	−4 / −16	−2 / −20	−9 / −17	−8 / −20	−12 / −20	−11 / −23	−15 / −33	−16 / −24	−15 / −27	—	—	−19 / −31
6	10	−3 / −12	0 / −15	+1 / −21	−7 / −16	−4 / −19	−3 / −25	−12 / −21	−9 / −24	−16 / −25	−13 / −28	−19 / −41	−20 / −29	−17 / −32	—	—	−22 / −37
10	14	−4 / −15	0 / −18	+2 / −25	−9 / −20	−5 / −23	−3 / −30	−15 / −26	−11 / −29	−20 / −31	−16 / −34	−23 / −50	−25 / −36	−21 / −39	—	—	−26 / −44
14	18	−4 / −15	0 / −18	+2 / −25	−9 / −20	−5 / −23	−3 / −30	−15 / −26	−11 / −29	−20 / −31	−16 / −34	−23 / −50	−25 / −36	−21 / −39	—	—	−26 / −44
18	24	−4 / −17	0 / −21	+4 / −29	−11 / −24	−7 / −28	−3 / −36	−18 / −31	−14 / −35	−24 / −37	−20 / −41	−28 / −61	−31 / −41	−27 / −48	—	—	−33 / −54
24	30	−4 / −17	0 / −21	+4 / −29	−11 / −24	−7 / −28	−3 / −36	−18 / −31	−14 / −35	−24 / −37	−20 / −41	−28 / −61	−31 / −41	−27 / −48	−37 / −50	−33 / −54	−40 / −61
30	40	−4 / −20	0 / −25	+5 / −34	−12 / −28	−8 / −33	−3 / −42	−21 / −37	−17 / −42	−29 / −45	−25 / −50	−34 / −73	−38 / −54	−34 / −59	−43 / −59	−39 / −64	−51 / −76
40	50	−4 / −20	0 / −25	+5 / −34	−12 / −28	−8 / −33	−3 / −42	−21 / −37	−17 / −42	−29 / −45	−25 / −50	−34 / −73	−38 / −54	−34 / −59	−49 / −65	−45 / −70	−61 / −86
50	65	−5 / −24	0 / −30	+5 / −41	−14 / −33	−9 / −39	−4 / −50	−26 / −45	−21 / −51	−35 / −54	−30 / −60	−41 / −87	−47 / −66	−42 / −72	−60 / −79	−55 / −85	−76 / −106
65	80	−5 / −24	0 / −30	+5 / −41	−14 / −33	−9 / −39	−4 / −50	−26 / −45	−21 / −51	−37 / −56	−32 / −62	−43 / −89	−53 / −72	−48 / −78	−69 / −88	−64 / −94	−91 / −121
80	100	−6 / −28	0 / −35	+6 / −55	−16 / −38	−10 / −45	−4 / −58	−30 / −52	−24 / −59	−44 / −66	−38 / −73	−51 / −105	−64 / −86	−58 / −93	−84 / −106	−78 / −113	−111 / −145
100	120	−6 / −28	0 / −35	+6 / −55	−16 / −38	−10 / −45	−4 / −58	−30 / −52	−24 / −59	−47 / −69	−41 / −76	−54 / −108	−72 / −94	−66 / −101	−97 / −119	−91 / −126	−131 / −166
120	140	−8 / −33	0 / −40	+8 / −55	−20 / −45	−12 / −52	−4 / −67	−36 / −61	−28 / −68	−56 / −81	−48 / −88	−63 / −126	−85 / −110	−77 / −117	−115 / −140	−107 / −147	−156 / −195
140	160	−8 / −33	0 / −40	+8 / −55	−20 / −45	−12 / −52	−4 / −67	−36 / −61	−28 / −68	−58 / −83	−50 / −90	−65 / −128	−93 / −118	−85 / −125	−127 / −152	−119 / −159	−175 / −215
160	180	−8 / −33	0 / −40	+8 / −55	−20 / −45	−12 / −52	−4 / −67	−36 / −61	−28 / −68	−61 / −86	−53 / −93	−68 / −131	−101 / −126	−93 / −133	−139 / −164	−131 / −171	−195 / −235
180	200	−8 / −37	0 / −46	+9 / −63	−22 / −51	−14 / −60	−5 / −77	−41 / −70	−33 / −79	−68 / −97	−60 / −106	−77 / −149	−113 / −142	−105 / −151	−157 / −186	−149 / −195	−219 / −265
200	225	−8 / −37	0 / −46	+9 / −63	−22 / −51	−14 / −60	−5 / −77	−41 / −70	−33 / −79	−71 / −100	−63 / −109	−80 / −152	−121 / −150	−113 / −159	−171 / −200	−163 / −209	−241 / −287
225	250	−8 / −37	0 / −46	+9 / −63	−22 / −51	−14 / −60	−5 / −77	−41 / −70	−33 / −79	−75 / −104	−67 / −113	−84 / −156	−131 / −160	−123 / −169	−187 / −216	−179 / −225	−267 / −313
250	280	−9 / −41	0 / −52	+9 / −72	−25 / −57	−14 / −66	−5 / −86	−47 / −79	−36 / −88	−85 / −117	−74 / −126	−94 / −175	−149 / −181	−138 / −190	−209 / −241	−198 / −250	−295 / −347
280	315	−9 / −41	0 / −52	+9 / −72	−25 / −57	−14 / −66	−5 / −86	−47 / −79	−36 / −88	−89 / −121	−78 / −130	−98 / −179	−161 / −193	−150 / −202	−231 / −263	−220 / −272	−330 / −382
315	355	−10 / −46	0 / −57	+11 / −78	−26 / −62	−16 / −73	−5 / −94	−51 / −87	−41 / −98	−97 / −133	−87 / −144	−108 / −197	−179 / −215	−169 / −226	−257 / −293	−247 / −304	−369 / −426
355	400	−10 / −46	0 / −57	+11 / −78	−26 / −62	−16 / −73	−5 / −94	−51 / −87	−41 / −98	−103 / −139	−93 / −150	−114 / −203	−197 / −233	−187 / −244	−283 / −319	−273 / −330	−414 / −471
400	450	−10 / −50	0 / −63	+11 / −86	−27 / −67	−17 / −80	−6 / −106	−55 / −95	−45 / −108	−113 / −153	−103 / −166	−126 / −223	−219 / −259	−209 / −272	−317 / −357	−307 / −370	−467 / −530
450	500	−10 / −50	0 / −63	+11 / −86	−27 / −67	−17 / −80	−6 / −106	−55 / −95	−45 / −108	−119 / −159	−109 / −172	−132 / −229	−239 / −279	−229 / −292	−347 / −387	−337 / −400	−517 / −580